Away from Tipperary, Nicholas Sadleir, Australian Gentleman

Away from Tipperary, Nicholas Sadleir, Australian Gentleman

by Robert Hodge

© copyright Robert Hodge 2014

This book is licensed for your personal enjoyment only. This book is sold subject to the condition that it shall not, by way of trade otherwise, be resold, hired out, or otherwise circulated without the author's prior consent.

Contents

Preface and Acknowledgements	vii
1. A Long Way from Tipperary	1
2. To Australia	25
3. Victoria and Gold	43
4. The Riverina	65
5. Head Stockman	99
6. Managing Albemarle	127
7. Anna Georgina and Quamby	167
8. The Quamby Years	201
9. The Adelaide Years	239
10. Ruin and Drought	251
11. Myths and Remains	273
	291
Bibliography	293
About the Author	295

Preface and Acknowledgements

This is about a set of great-grandparents – Anna and Nicholas Sadleir, and their families, who made their mark on the world in the 19th and 20th centuries. It uses public records as a framework and links major and minor events with family anecdotes and fantasies about what happened in their lives. People in the story with the surnames: Abbott, Barry, Bell, Brooke, Brush, Burke, Carter, Chadwick, Clarke, Crofton, Cromwell, Crowe, Cuddeford, , Dry, Dodery, Falkiner, Flood, Fox, Furage, Gladstone, Hamilton, Hannaford, Hunt, Kelly, Kidman, Lawson, Lord, McDonald, Patterson, Parker, Payne Peterswald, Phelps, Ritchie, Sadleir, Raimond, Sams, Sargent, Scharpel, Scott, Spence, Sturgess, Tapscott, Turpin, Urquhart, Vandeleur, Waddell, Westbrook, Wilson or Wills lived in Argentina, Australia, Ireland, England, New Zealand, South Africa or Scotland, or several of those places. Others may have existed but perhaps not using the names I gave them. All in the story lived authentically for the times – it was my job to make them real. I hope it worked.

I have cousins to thank. Barbara Stacy of Adelaide started this adventure by asking for details of my children and grandchildren for a family tree she was making. She stimulated me with stories about my great-great-uncle's arrest of Ned Kelly, the book he wrote, reports of family nobility, stories of her father and his brothers and sisters, and with photographs and papers. An 1882 diary came from her nephew James. Barbara lent me her copy of a book by Claudia Richards Mousely. Claudia was from the family who had the homestead section of a sheep station Nicholas managed and she wrote several chapters about him and his times. Claudia helped me with data and references.

Barbara introduced me to Richard Sadleir of New Zealand, a great-grandson of Nicholas's brother John. Richard had written an authoritative genealogy of the Sadleirs and he stimulated and encouraged me with newspaper references to Nicholas' pastoral holdings and his evidence to government enquiries. We had some fine debates about significant events.

Ronnie Land of Glasgow, Scotland, had huge genealogies containing the ancestors of our great-grandparents, and he wryly concluded that he and I were 14th cousins of the late Diana, Princess of Wales. He and his wife Maureen hosted me in Glasgow. Ronnie told me the story of his mother whose father, Robert Sadleir, Anna and Nicholas' fourth son, settled in Patagonia and sent her to boarding school in Scotland where she remained and married. He sent me home with hundreds of files.

There were others: Dennis Murnane, a Tipperary historian, helped with accounts of Sadleirs, the Tipperary violence of the 1840s and the potato famine. Unnamed librarians at the National University of Ireland, Maynooth, the Thurles library, the Kings Inns library and the Quaker Historical Archive and Library in Dublin helped with research on the Sadleir and Phelps families in Tipperary, Limerick and Clare counties. Newly discovered Tipperary cousins, Patrick and Edel Merrigan, told me about Marshal Sadleir, the last Sadleir owner of Brookville House near Tipperary, and Conor Crowe, a son of the present owner, showed me through it.

Australian research was less formal, but I received gracious attention from the staff and management at Quamby in Tasmania, now a holiday resort, and from Nicholas Klemm and his daughter Sharon Bonsalaar and her husband Nicolaas at the old Albemarle homestead called Windalle. The owners of Bingara near Eulo in Queensland guided me to Fitzherbert Brooke's grave and were helpful with the history of that country. Help came from museum staff at Longreach, Winton, Wentworth, Westbury and Cloncurry and from Shire Council staff at Boulia and Mount Isa. I spent days in the New South Wales and Queensland State Archives, the Holdfast Bay History Centre found houses Anna and Nicholas and the children used in Adelaide and the University of Adelaide helped with records of some of the children.

Most information came from the World Wide Web where I stumbled over seemingly unconnected pieces of information and worked to meld them. And I spent time listening to people in pubs in Ireland and Australia.

To everyone who helped me or made me do things better, a sincere thank you.

1

A Long Way from Tipperary

The author, a retired 70-year-old, starts a quest for the story of his great-grandparents on the Birdsville Track by talking to his great-grandfather who has been dead for 107 years. It is the best season for 50 years. They rejoice in it as the great-grandfather describes his drovers grazing the channel country as they walked cattle for months to the railhead at Marree for shipment to Adelaide. They talk of sheep and cattle runs the great-grandfather had. The great-grandfather talks of his privileged childhood in Ireland, insurrection and the potato famine before he leaves with his brothers for Australia.

Mungerannie, August 2011 – I sat in my car parked on gravelly clay outside the hotel opposite the fuel pumps talking to my great-grandfather. Mungerannie was on the Birdsville track between Marree in South Australia and Birdsville in Queensland. The place was a waterhole on an abandoned desert stock route where cattle walked from Queensland cattle runs for months to the railhead at Marree for rail transport to Adelaide. Only tourists, geologists and stock transporters use it now. The pub was a commercial gamble and my great-grandfather, who had been dead for 107 years, said it was when he was alive too. He told me about Mungerannie:

'We walked the stock route to the railhead at Marree for several drives of cattle we sold in Adelaide. The South Australian government sank a bore here, if I remember correctly in about 1900, but there was a well here long before that, with a hotel of sorts and our drovers used to water our cattle here on the way down from Queensland.'

Pam rapped on the window. She was smiling at me tentatively – middle-aged – she had the friendly and tolerant look of someone who was used to motorists chatting to a windscreen.

'Are you okay? Can I help you?'

I started, blushed, got out of the car, looked at her with a silly grin, looked at the ground and muttered.

'Sorry. I was sort of talking to myself. Yes. I hope you can help me.'

I asked her humbly for a room, a meal, and said, almost casually, 'I have a small hole in my fuel tank and I wondered if anyone here could help me fix it.'

'Well there's no problem with the room or the meal, Phil's out the back, if you can wait five minutes, I'll get him to have a look at the problem, he's pretty handy.'

Great-grandfather Nicholas Sadleir remained silent. I imagined him smiling. His awkward 70 year old great-grandson, daydreaming and ill-prepared, had been rushing about and getting into trouble in Ireland, New South Wales, Queensland, Tasmania, Victoria and now in South Australia. He spoke in a familiar way to unreliable people of uncertain political stances and he seemed to believe what they told him – Australians, it seemed, had become more ill-mannered and independently spirited. The questions about Sadleirs in Ireland and Australia made Nicholas Sadleir reflect on his life and times – his 15 children and the fortunes he had and lost, but it was strange to be talking to a great-grandson who was older in years, had lived in the luxury of the 20th and 21st century (he even had a motor car with a telephone and interrupted conversations by taking calls on it and used it to photograph a couple of dingos on the track) and asked so many questions about Ireland and Australia in the 19th century. The great-grandson spent extravagantly.

Dingos on the Birdsville Track

He posed questions about blackfellows, money, marriage, crime, affection, friendships, politics, nobility and class, law keeping and religion in Australia and Ireland. His journey down the Birdsville track served no purpose. The track was for drovers with cattle, or mailmen, not for lone motorists. What was the point? He, Nicholas, had never travelled south of Boulia on it. There was a family to care for, stock to buy and sell and stations to manage.

Nicholas Sadleir wasn't there. Neither were his siblings who came to Australia: Richard, a Melbourne surgeon, Marshal, a Mansfield lawyer, famous John, the policeman who supervised Ned Kelly's capture at the siege of Glenrowan, nor Nicholas' twin, Helena, who vanished. I'd imagined them from stories about them and the history of the times they lived in. I thought they were noblemen. History, photographs and imagination drove me. He and his brothers and sister had been real. I wanted him real again but there was nothing spiritual about it. He simply made a good travelling companion. With 15 children, he had to have been a reasonable parent. That made him a useful great-grandfather and storyteller. We were on our way south from cattle stations he had in Queensland. The country was looking wonderful. It had had good rains for two years. Before we got to Mungerannie we'd got to know each other better. Nicholas knew I was his daughter, Georgina's, grandson. I'd told him that when I'd started the conversations we had on the way to Queensland but we were awkward with the way we talked. He called me Robbie at first to not confuse me with his son Robert and I called him Great-Grandfather.

'Robbie, it seems we are getting to know each other. Great-grandfather seems too formal, and, in the scheme of things, you are my senior. I died when I was 68 and you are 70. You address me as an old man and I address you as a child. What do you your friends call you? '

'Red.'

'Why.'

'Because I had red hair.'

'Had? Are you grey now? I went grey in my forties, but I had a redheaded daughter.'

'No. It's still red.'

'Why aren't you called *Blue*. I had several redheaded coves we called *Blue* on Albemarle. We had a Menindee *Blue,* a Booligal *Blue* and a Victoria Lake *Blue*. They were reliable coves although one was a bit quick-tempered and spent a lot of time in the Wilcannia lock-up after a spree whenever he went to town.'

'When I went to boarding school there was already a *Blue* there before me so they called me *Red,* Great-Grandfather.'

'Well I shall call you Blue henceforth, but please desist from calling me Great-Grandfather. Call me Nicholas.'

"No. If you choose to call me *Blue*, you will be *Holas*."

"*Holas!* I've never been called that."

"And I've never been called *Blue.*"

'A hard bargain Blue.'

'Yes but Holas isn't a bad name. It's dramatic with its oratorical beginning, dignified and soft-sounding and it gets rid of *Nic* – the devil in you – if there is any. I believe you were something of a church administrator in Tasmania. To support that, Holas sounds holy. And it's ideal for somebody who's dead because there isn't a living soul I know who answers to that name – so nobody will get mixed up and answer for you when I'm talking to you.'

'You're planning a long conversation, Blue?'

'I reckon you may have a hell of a story, Holas.'

'What do you want to know Blue?'

'Everything. But to begin, how did you learn enough to manage one of the biggest sheep stations in Australia less than 10 years after your arrival from Ireland as a raw teenager?'

'Teenager, Blue?'

'It means someone between 13 and 19, Holas. It's the TEEN in the word that helps to classify them – someone moving to adulthood from childhood. Did you not call them that?'

'Never.'

'Well what did you call them, your children, when they were at that stage?'

'Youth! Blue. We called them Youth. We spoke the Queen's English.'

'Yes well I suppose you did at home Holas. But what about in the goldfields or the stock camps?'

'Yes well – perhaps that was another matter – but no one ever called anyone a TEENAGER'.

'Well what did you call yourself when you arrived in the colony? Surely you didn't call yourself a youth, Holas. Did you call yourself a young gentleman?'

'Well yes.'

'You're bloody joking.'

'Usually I left out the *young*.

'Did you call yourself that all your life?'

'It became a slightly more awkward to use it by the time of the federation of the colonies. People wanted to seem slightly more equal, and we didn't normally use it in conversation – it came up mainly in correspondence or newspaper reports. I was often called Nicholas Sadleir, Gentleman. Other people used it. One didn't usually denote oneself a gentleman even if one was.'

'Nowadays being a gentleman can literally mean keeping one's unwelcome hands off females – or, with more refinement, good manners, opening doors, making people feel less shy in new circumstances, and so on. I think it has changed a bit. In your day Holas, it was a rank, a position even an obligation. It placed you as a wealthy man who did no work?'

'Yes. You described it correctly, Bluey. Gentlemen worked, but not manually. They directed and planned and invested. They commissioned professionals. They led good order. That sort of thing.'

'Were you ever called a shearer, Holas?'

'Never, Bluey.'

'Why not?'

'Because I wasn't. In its own way it denoted an occupation and standing – just like gentlemen.

'I was in hospital with a shearer, who was laid up with pleurisy because he had been shearing wet sheep, and he read out a newspaper report saying that *two men and a shearer were involved in a serious road accident at a rail crossing.* He was furious that the shearer had not been awarded the rank of *man* by the newspaper's editor.'

'Blue, that ranking I can understand.'

We were near Boulia in Queensland going to South Australia when we agreed on our names.

'Was it as good as this when you were here, Holas?'

'I can't see what you can see, Blue.'

'I'll describe it. I think this is the best the country has looked for at least 50 years. It's rolling Mitchell grass plains. The trees are sparse; we're getting close to the Channel Country. The cattle are sleek, fat and shiny. There is probably enough feed for five times the stock that is here.'

'But it doesn't last.'

I knew that. In a year or two, the cattle would be leaner and the ground barer. This plenty wasn't normal, austerity was. It was why I had come this way. I was unlikely to see seasons like this again.

He expanded. 'I managed Albemarle on the West Darling in New South Wales from 1862 to 1904, and I'd been in the district for four years before that. We kept good records. In this arid back country there is no such thing as a normal year. We had droughts from 1864 to 66, again in 1868, again in 1877 and then good years before a run of poor seasons from 1882 to 1886. We had a terrible drought in 1889 but the centenary droughts continued for about four years. That finished me. In the early 1860s we had bounteous years and we ran up to 200,000 sheep. Then back to 75,000 in the 1880s. The worst run was from 1898 to my last years. We got down to less than 5,000.

I can't see this country now, but even if you described it, I can't compare it with my memory of it. I never took this track south to Bedourie and Birdsville but we certainly knew about this Channel Country. This was the fattening country we could use but not own. It often bloomed after a flood when our country further north was dry so we put stock on the road with drovers. Sometimes we sold them in Adelaide, and occasionally, if we had good rains back at Cloncurry or on the Templeton, we would get word to our drovers to turn our herds around and bring them home. They were often on the road for six months.'

Water from here was on its way to South Australia. When the rivers ran (they were more often dry than wet) they flowed inland to Lake Eyre. But mostly Lake Eyre was barren with a salty crust. The inland rivers didn't usually get to it – they filled billabongs and lagoons along the way and petered out. Lake Eyre had filled only four times in my lifetime, and it had flooded last year and would get Queensland water from more than 1000 miles away again this year

Map of the Lake Eyre Basin

As we went south the country changed. Tall red sandhills governed the course of the road and the streams and waterholes beside it had pelicans. I wound through, over and beside the sandhills. Most of this country had not flooded and it looked parched. I pictured turbaned Afghans leading groaning camel strings plodding beside sandhills carrying bundled sheets of galvanised iron on either side of their humps for buildings on stations.

I pictured turbaned Afghans leading groaning camel strings plodding beside sandhills carrying bundled sheets of galvanised iron on either side of their humps for buildings on stations.

'Did you use camels, Holas?'

'Oh yes, Blue – for Albemarle and Bingara. There were Afghan families in Broken Hill with camels and they helped us cart wool when the paddle steamers sat in a dry river when we needed to get wool to the Adelaide auctions. The camels carried two bales each (some of the big camels carried four) from Albemarle to Broken Hill and the wool went by train down to Adelaide for sale. And of course we used contractors with camels to take things out from the paddle steamers to parts of the station away from the river. We carted coils of wire out to fencing contractors that way. Some of our back paddocks were more than 50 miles from the river.

Camels carted stores and fencing wire a couple of times from the wharves at Wilcannia to our cattle station Bingara east of here, close to Eulo, near the New South Wales-Queensland border. That was a big trip. Probably more than 300 miles. And remember, most of the goods came from England via Melbourne – by train to Echuca – then on to paddle steamers going down the Murray River to Wentworth and then up the Darling to Wilcannia. Some of those coils of wire might have been travelling for nearly a year!'

More broad, long sandhills, less waterholes as the country became more desert-like as it led to Birdsville. Overall, the country ran flat but sandhills sometimes made it mountainous. There were fewer trees – sparse stunted shrubs instead. I saw no one.

I reached Birdsville in four hours from Bedourie. An aeroplane had landed from Brisbane with the mail and the pilot was having lunch at the pub before he continued his round.

'It's a milk-round mate. We service Boulia, Bedourie, Mt Isa, Charleville, Quilpie and Windorah and we connect with Brisbane. It's a good service. This is a day/night airstrip so we can get in and out reliably.'

Considering the size of the town (about 20 buildings) and the population of the district it supported (probably less than 200) Air Atlanta Icelandic gave luxurious service.

I wondered why an Icelandic airline was flying in outback Queensland and the pilot shrugged. 'Dunno mate, it probably just made commercial sense at the time. The whole thing is more or less an Australian operation; it just works under the banner of Air Atlanta Icelandic. We run a service to most of inland Queensland, it's a more or less regular service – a sort of cross between that and a charter flight – if there is a mail run, we have a regular flight but on the others, if there are not enough bookings, we don't go. Passengers may have to wait a day or so.'

'Have you anything to say about Birdsville, Holas?'

'Hardly a fair question, Blue. As you know, I've not been here, but in my day fellows called it a wild and dangerous place. There was always a pub for drovers to get into trouble, and there was a police and customs post to maintain some semblance of order. We had to pay duty on goods passing into South Australia, and I can remember one of our drovers complaining about his bags being searched in case he was smuggling Queensland rum or Chinaman's opium. Strange! Of all the drovers we had he was the only teetotaller.'

The Diamantina River south of Birdsville had receded enough to let me cross it on the road south to Marree – the famous Birdsville track – 'the loneliest track in the world'. It had been officially 'open' for a week or so. It was flooded for weeks before that. This was the 'outside track'. The 'inside track' was shorter but it would be closed for months. Goyder's Lagoon flooded it with water from Queensland's rivers.

The sandhills grew taller as I drove in dust beside them. I crossed into South Australia but the country stayed beautifully harsh. It had proper roads only in the last 40 years. Before then, motorists carried sheets of metal to lay a temporary road to get them over sandhills. Often they made less than 50 miles a day.

At first, the mail service to Birdsville came from Marree in South Australia. Mail contractors used packhorses and camels or horses hauling buggies and stagecoaches – depending on the track. Entrepreneurs tendered for the Royal Mail contract (usually, the cheapest bid got the job) but they made money out of goods and passengers too. The mail contract formed the skeleton and freight put on flesh.

As I headed south a rock the size of a watermelon crashed into the fuel tank leaving a split and a dribble of diesel. I looked, knew I couldn't mend it, calculated the distance to Mungerannie Bore, tried to guess the fuel I was leaking, declined to speak to my great-grandfather about it, leapt behind the wheel and sped to this halfway hotel.

And so Holas and I talked outside the Mungerannie pub until Pam stopped us and Phil mended the split in the fuel tank. It was the end of a travelling quest.

Nicholas Clarke Sadleir and Anna Georgina Sadleir were talented Australian colonists. I'd learned about them and their children in Ireland, Australia and Scotland. Sadleir brothers came from Tipperary in Ireland following the potato famine. They prospered in Australia. They had large families. Descendants live in Argentina, Australia, England New Zealand and Scotland. Some had been, or nearly became, wealthy members of the British aristocracy.

It was time to tell their tales from the beginning.

There were many Australians looking for ancestors in Ireland in 2010. I was one of them. I had just left Brookville House, south of Tipperary town. Nicholas Sadleir was born there. There were no

Sadleirs there now. The Crowe family had it. Mr Crowe bought it from the estate of Marshal Sadleir in 1964.

I read this aloud:

Containing 117 acres 0 roods, two perches or thereabouts,

This most desirable property is situate on the Road to Glen of Arherlow, within one and a half miles of Tipperary town, it is well served for marts, creameries and all other amenities.

The lands are excellent limestone quality, all under pasture, entire without waste, well fenced and sheltered and have a never failing supply of water from springs, streams and the River Ara, which forms part of the boundary.

There is a very substantial Georgian residence, approached by a short avenue, and the accommodation comprises large hall, large dining room, study, two reception rooms, kitchen, scullery, and hot room and bathroom on the ground floor; large landing, six large bedrooms, one smaller bedroom and bathroom on the first floor, French window in reception room leads to a wall in the kitchen garden and orchard at the rear. The outbuildings contained in two independent yards and are all stone built and slated. Yard number one has independent entrance from the road and contains a cow house (partially lofted) to tie 40 cows; machinery house (lofted); dairy; two standing stalls; harness room (lofted): feeding house, fuel house and four column hay barn. Yard number 2 which is attached to yard number 1 contains large barn and fuel house; storehouses: large garage and machine house (which is the only building covered with iron). Galtee supply laid on to residences, yards and concrete tank in lands E. S. B. Installed throughout.

There is also a two-storey residence (in need of repair) which will be sold as a separate lot.

The special attention of those in quest of the most outstanding and attractive dairying or fattening holding is directed to the sale of this very valuable property, or to those requiring an ideal hunting residence being situated in the centre of the Scarteen and Tipperary hunting country.

'Can you hear me, Holas? Do you know what I'm describing?'

'I believe I do. Where does the description come from? Could it be my old home in Tipperary?'

'It's from a Tipperary newspaper, an advertisement to sell Brookville House from the estate of your nephew Marshal Sadleir. He was your oldest brother James' son.'

I waited.

His voice had changed. 'Yes. We had news of Marshal's birth. We got letters from home. But have you seen the house, Blue? Are you there? Please tell me what you observe?'

'I'm at the end of the drive leading to it on the edge of the road south of Tipperary town. I can see the house and I've just been in it. It's a large two-storey house with an archway at the side leading into yards, stables and farm sheds. It has an elegant reception hall with a skylight and a staircase, large dining room to the left, and a sitting room or salon – perhaps you called it a drawing room – to the right. A large kitchen leads off the dining room and there is an entrance to the kitchen coming from the side yard as well. External stairs lead to the front door which has an ornamental glass arch above it. How does that serve?'

Brookville House, South of Tipperary Town.

'Well done, Blue. I remember it well from your description. I left this house for the colony of Victoria when I was 17, in 1852 but I had a lovely childhood. Respectable Christian parents. Loving household staff and a tutor before I went off to school at Midleton College in County Cork. Children from tenant farms to play with. Older brothers, a twin sister. Grandparents, aunts, uncles, and cousins nearby. And friends from families of the other big houses for church, levees, hunting, horse racing, balls, serious discussion and deeper friendships. [1]

I didn't always think so then, but we had a good life. Sadly, I never repatriated. I wanted to, but I never contrived to do it.'

'What is your first memory at Brookville house, Holas?'

I waited.

'It was a funeral. Our mother and father left us in tears in the care of our nanny to go to it and my sister Helena pleaded with our mother to be allowed to go. At first, I sided with my mother.

"Girls may not go to funerals," I said.

"Nor may boys," my mother said and she started weeping too.

I must have other memories, but it was the first time I had seen my mother cry. I forgot about it quickly because Cook and Nanny bribed us with apple pudding as my mother and father drove off with coachman Liam in the carriage in their black clothes, but I connected with the story later in life because it was the funeral of my mother's brother, Uncle Patrick Clarke. Some of his tenant farmers shot and bayoneted him. It caused an outrage at the time; two people were hanged and another escaped to America. I think my mother grieved for him all her life. She was very fond of him, and he was kind to us, but all sister Helena and I thought of at the end of the day was

---

1. James Sadleir (born 1792, died 1867) married Elizabeth Hare Clarke (born 1799, died 1889). They lived at Brookville House, Tipperary and had the following children: James Robert Sadleir(born 1820) Richard Sadleir (born 1822, died 1822, South Yarra, Victoria, Australia. Alicia Sadleir (born 1825) Marshal Clarke Sadleir (born 1827, died 1903, Mansfield, Victoria, Australia. Robert Sadleir (born 1831, presumably died as an infant) John Sadleir (born 1833, died Melbourne Victoria 1919) Nicholas Clarke Sadleir (born 1835 as a twin to Helena, died 1904, Menindee, NSW, Australia) Helena Sadleir (born 1835, there is some indication she died in Australia) Elizabeth Bolton Sadleir (born 1845, died 1872, Brookville) Mary. E. Sadleir (born 1862, died 1950, Brookville)

the apple pudding. I can still taste it. Mrs Ryan was an accomplished cook.

We had fine horses to ride – and we raced them with lighter jockeys in local meetings. Brothers James, Marshal, Richard, John and I joined in with the local hunt club when we could. I used to play football with some of the tenant family boys, but Pater didn't encourage fraternisation so I stopped when I went to Midleton College. We played English games there. I enjoyed tennis and cricket and we had a few games on stations in Australia – I was a host at tennis competitions in Northern Tasmania.

The next thing striking my memory was my getting ready to board the stagecoach in Tipperary town to go to Midleton when I started boarding school. I suppose I was about 13. Helena was weeping. She was my twin. I think because we were babies of the family (there was a younger sister but she was tiny), the brothers, the servants, and even Mater and Pater, allowed us more comforting affection than would normally have been allowed between brothers and sisters.

Helena was angry too. "You are going off to Midleton, Nicholas, just so I will have to do all of the Greek and Latin verb conjugations by myself. And who will pick all the flowers for Mrs Nicholson or look to the hounds?"

We were in the hall. Mater held out her arms to comfort her, but brother Richard, who was down from Dublin between courses at Trinity College, started to tease her: "Little Hellie, the gardener's maid doesn't know what will become of her." Helena ran up the stairs furiously. I looked to my father.

"Best you just go now, Nicholas. Leave her. She will be better soon." And he glared at Richard.

Liam took me in the buggy and pair to the coaching station in town. It was raining and we both wore hats and heavy woollen coats. We didn't utter a word on the way. The coach was waiting in the cobbled main street. Liam tossed my bag to the coachman who strapped my trunk atop the coach with leather ties. "This young gentleman is for Midleton via Cork. Have a care he gets there safe." And he got down, turned to me, took my hand and said "Good luck,

Master Nicholas. We will look after Helena. I'm sure you will do the family proud."

So off I went in the dripping coach. Brother John was already there as a pupil. He was waiting for me. It was evening and I connected to the Dublin mail coach and travelled all night to Cork city in the rain. We had four stops for refreshments and to change horses and I had to wait in Cork for half a day for a coach out to Midleton. Two ruffians tried to rob me twice but I drove them off while I was waiting at the coaching station in Cork city and I remember being admonished for being late by an under-master when I walked to the college with a barrow boy pushing my travelling trunk from the coaching station, but John came to my aid. I had an interview with the headmaster, the Reverend Turpin, a plain dinner in the long noisy refectory with the rest of the boys and masters, settled into my dormitory and slept all night. I was exhausted.'

'And did you do the family proud, Holas? Cups? Medallions? That sort of thing?'

'I was a fair scholar Blue, but John was the outstanding sportsman. I stayed for two or three years. I expect I conformed to the standards set by my brothers – and please don't laugh, Blue – I won an exhibition for my scholarship in Divinity.

'I'm not laughing. I got one too.'

'Really! You're a distinguished theological scholar too, Bluey?'

'No. I reckon the school chaplain wanted to encourage me in continuing life so he changed the marks. I hated boarding school. Did you Holas?'

'No. I really was good at remembering and reciting the scriptures so I didn't need to be bribed to continue. In the end I had to conceal my talent. The family aspired to holy orders for me. I didn't seek life as a clergyman! Australia saved me. Perhaps it would have been better for you Blue.'

I declined to answer. He went on. 'Like me, Marshal and Richard qualified for entrance to Trinity College, Dublin, but only Richard attended as part of his medical studies. Marshal was an apprentice attorney with our uncle Nicholas in Dublin – he qualified as a lawyer without going to Trinity. John was an excellent sportsman, but he didn't take the entrance examinations – I think he looked to a career in the military. I qualified for Trinity College entrance but I never enrolled. Fortunes had changed in the family. Money for fees was scarce.'

'What happened?'

'Surely you know of the potato famine, Blue?'[2]

'I know a bit. A disease killed the crop for several years in a row. But tell me more; I had always imagined the Sadleirs in Ireland with independent incomes.'

'We had been comfortable until the potato famine and a few of the other branches of the family could be called wealthy, but the potato famine ruined everyone's livelihood – from the highest to the lowest. Tenants lost their land, labourers lost employment, and some of the large estates were broken up and sold.'[3]

'How was it for you and the rest of the family?'

'We had sufficient, and we gave food to our tenants and other local families in difficulty, but it was terrible for many ordinary people. When I was at Midleton College I saw dead people on the side of the road with green slime coming out of their mouths. They'd been eating grass. The college always had enough food for the masters, pupils and servants. At first the housekeeper and cooks armed themselves to guard the food, but the headmaster intervened and from then on a line of people waited in the kitchen yard for scraps from about midday onwards. Dr Turpin espoused and practiced Christian charity.'

'Did you go without?'

'No. But we were made to think that we were sacrificing

---

2. Thousands of people died. A "Poor Law Union" operated. Owners and occupiers of land paid tax for relief for the poor. In 1845, the famine exhausted reserves. One third of people starved. Tipperary's poor law union could not cope. It dissolved and the central government from Dublin took over. More than 21,000 people relied on it as thousands died.

3. Between 1849 and 1856, landlords evicted 22,000 people from South Tipperary. Emigration saved many. For Ireland, 75,000 left in 1845 increasing to 250,000 by 1851. About 18,000 people left County Tipperary then. And some of the "Ascendancy" went broke. Some landed families lived extravagantly and the famine was "the last straw." Their debts exceeded income. In 1849 Dublin Castle set up the Incumbent Estates Court to allow rapid land sales.

something for the good of the poor. And we prayed for them a lot, and there were many sermons about the need for Christian charity and care.'

'Why did you go to Midleton College? Wasn't your grandfather, Marshal Clarke, headmaster of the Abbey School at Tipperary?'

'Yes he had been a headmaster and clergyman, but at about the time we boys needed a school (we had a tutor at home) the army purloined the school buildings to use as barracks.'

'Why?'

'Politics, Blue. Rebellions were starting. I've never had much time for the Fenians. Probably one of the benefits of the potato famine (it is probably unwise of me to promote this opinion) is that it extinguished many of the rebellious tendencies held by supporters of the Papacy and Irish independence. Not to joke about it – but they lost their stomach for it, and many of their leaders left for America or Australia (I met several there and they had started a new and reformed life). But I never paid much attention to the detail, Pater and Marshal debated it vigorously, there were reading rooms in town, but I could not maintain an interest.'[4]

'I was probably too young when the rebels were most active.[5] Brother John told me about rifle balls coming through our front window when we were small boys, and one of our cows was shot, but John didn't think this had much to do with the general revolution or religion. It was about threats to Pater who was giving evidence against somebody who had assaulted someone on a road to a market. Pater never mentioned it, and John told me about it after we had arrived in Australia.'

---

4. There were three 'news and reading rooms' – a Clanwilliam club for Tories, the Repeal room for O'Connellites, and the moderate Subscription room. Daniel O'Connell championed Roman Catholics – 85% of the people. He persuaded the British to let Roman Catholics enter Parliament in 1841. He opposed tithe taxes for the Church of Ireland (and nothing to the Roman Catholics). His 'monster meetings' for an Irish parliament made O'Connellite thinking part of the Irish psyche.
5. Before the famine, Tipperary people took violent action in disputes about farming land. Eighty-five per cent of the people had no votes. Tipperary residents killed 11 landlords, nine employees of landlords and eight farmers.

'Did your family think of itself as Irish, Holas?'

'I think we did. Sadleirs had been in the country for about 200 years. We sometimes denoted ourselves Anglo-Irish. The Sadleirs in Tipperary, Limerick and Cork counties came down from Colonel Thomas Sadleir who arrived from Stratford-upon-Avon in England with Oliver Cromwell to crush an Irish revolution and to outlaw Popery. He was granted land here but it wasn't extensive.[6] The Sadleirs didn't have titles like others of the ascendancy, and most of our land holdings were modest. Many family members were clergymen in the Church of Ireland and there were lots of military men and barristers and solicitors and magistrates. We were listed as gentry in government lists – the families had income from land and capital.'

'Why did you go to Australia? How did you know about the opportunities there?'

'Yes. Well there's the rub Blue. James, the eldest, was to assist with the estates and he was to inherit the lands. That was settled. It was normal. Richard, John and I had to think about our futures. Marshal was to become an apprentice attorney with Uncle Nicholas in Dublin. Pater and Mater encouraged us to contemplate the professions and callings (I think they guessed family investments and lands could no longer support us although they were too gentle to mention it). John and I hoped for bucolic pursuits but our hopes of being able to purchase or lease additional land and livestock in Ireland were fading. When a letter came from a cousin who had been in Australia for more than 10 years with his brother, investing in sheep and land, telling us his flocks exceeded 100,000 head we started to imagine ourselves immigrating. This cousin was from a Limerick branch of my mother's family. He was not a Papist but nor was he from the Church of Ireland. He was a member of the Society of Friends – a Quaker, and we considered him somewhat degenerate. In our conceit, we thought that if he succeeded we could surpass his efforts. And there was the gold in Victoria. There was news of it in

---

6. Colonel Thomas Sadleir got about 5,500 Irish acres (8,923 ordinary acres) in Tipperary for his work in Oliver Cromwell's invasion.

abundance in all the news sheets. Many of our friends planned to go and urged us to join them.

We started to prepare secretly. Kathleen, the upstairs maid was the first to catch us. "Master Nicholas, why ever are you storing pistols, saddlery and riding boots beneath your bed? To be sure are you after thinking of riding out with the Young Irelanders?"[7]

I blushed. I was fond of Kathleen, but she was much older than I. I knew she would be hurt by what I planned to do but I resolved to make a clean breast of it. "You must not be teasing me about the Young Irelanders. They are a frightful lot. They challenged our Queen and the good order of things here. And anyway, they lost and are dispersed. That equipment's for adventures in Australia, Kathleen."

"Never!" she said and collapsed on the bed staring at me. She rose and started pacing. "So it's come to this. Is gold more important than us? Do the master and mistress know of this? This seems to be a crime worthy of punishment by transportation[8] and you seem set on punishing yourself by arranging that very thing. I have two cousins transported there two years ago. They do not report happy prosperous times on government road gangs! Think on it, Nicholas. What of your friends and your mother and father?" She rushed off down the stairs towards the kitchen. She clearly thought it her duty to inform all members of the household of my imminent desertion.

Dinner that evening became an ordeal. We still maintained formal service – glittering glass, shining silver – everything coming to one's plate from serving dishes brought from the kitchen by servants – a formal and solemn ritual, and we wore our best clothes. Kathleen helped with service (our footman was away visiting his sick mother in Golden) and she kept glaring and smiling at me by turns. I faced the front and ate fitfully. After Kathleen cleared the dessert things, Pater relieved my tension.

"Nicholas. What's this I hear of your plans to go to Australia?"

---

7. An Irish nationalist movement whose uprising in Tipperary was defeated in 1848. Several of the gentlemen ringleaders were transported to Tasmania.
8. Transportation to Australia was a punishment second only to hanging. Many Irish were transported.

Brothers Richard and John started and stared at me. I flushed. I was not afraid of my father, but his tone was unusually confronting.

John responded before I could reply, "Nicholas is not acting alone, Pater. I plan to go as well and Richard is pondering the idea. It isn't just the gold (although that is immediately attractive) but you will remember the letter from Mater's cousin, John Phelps, with news of the flocks of sheep he has accumulated. It seems Australia is a continent of opportunity."

Mater frowned, looked to Kathleen, nodded and Kathleen withdrew. I sensed we were to be admonished but not in the company of servants – but that is not how events ensued.

"But Richard, you have not completed your medical studies. Please explain your thoughts," my mother said.

"John was right when he said I am not firmly decided, Mater, but I fear that the practice of medicine or surgery here in Tipperary or in Dublin or Waterford or Cork will not be well rewarded. People have little money to employ professional healers. The talk in the Medical School is that I may be able to profit from my skills more easily in the colony of Victoria. All reports are of profligate prosperity. Opportunities for medical men seem to be without limit on the goldfields. We as brothers also propose to care for each other if we undertake the adventure together."

My father looked to John. "And you John. What are your arguments?"

"I suggest respectfully, Pater; the family is not in a position to buy me a commission in the Suffolk Regiment in the way that we talked about as a possible career for me after my schooling at Midleton. There are several chaps who were keen on military commissions and have given up the idea as I have, thinking that there may be careers in colonial leadership in Australia. There are British regiments there and we hear that their numbers are depleted with men and officers deserting for the gold diggings."

"And Nicholas?" Pater turned to me.

"I want to make my fortune in gold or wool, Pater, return here and restore our family's fortunes and standing."

Pater and Mater smiled. They seemed to be indulging me.

"Noble aspirations, Nicholas. Your reports from Middleton

suggest that you should thrive in a world of scholarship – holy orders, or a life of letters, perhaps even the law. My brother Nicholas already plans to apprentice your brother Marshal in his chambers in Dublin. I could make representation to your Uncle Nicholas on your behalf?"

"Thank you Pater, I do not wish to appear ungrateful but I do not admire Dublin. It seems its main purpose is to harbour rogues and thieves."

"Yes. And I imagine you would be at home there, Nicholas". Pater's eyes twinkled and my brothers chuckled; only Mater and Sister Helena lowered their eyes.

"I am determined to go, Pater, but I do seek your permission and blessing." I turned to my mother: "and, Mater, may I seek your good offices as well?"

My mother slowly smiled. The tension dissipated. Only Helena stared stonily into the garden beyond the window facing her.

Our parents looked to each other. It was a serious sad look. Then they looked at us with gentle smiles. Pater said, "We will not stand in your way, any of you. We are pleased to have had the means to educate you to enhance your endeavours wherever you go. And Helena, look up, be joyful. One day you may feel inclined to visit your brothers in Australia." And the door burst open and Kathleen came into the room with a tray of wine glasses and a decanter of port. "Madam, I thought the family might be after needing this for a celebration." She must have been listening at the door.

Mater told me later that Kathleen had reported our early conversation and she responded to Kathleen by telling her she knew John, Richard and I were contemplating going to Australia, she and Pater approved, and they planned to tell us that evening. In many ways, it was a kind and intelligent household.

We had cleared the air. In hindsight, I think the finances of the household may have been far worse than we knew. Perhaps, with our going, Pater could see a way to keep the household and the lands.

For John it may have been the greatest wrench of all. The following day he took a coach to Sligo to visit his sweetheart, Isabella Crofton, at the village of Skreen. Isabella's father was the Rector. John was away for more than a week and he returned looking sad. He said nothing to me but steadfastly planned to get to Liverpool and on a ship to Port Phillip Bay. He never told me he and Isabella planned to marry later (perhaps they didn't then) but she came to Australia and married John five years after he immigrated.

Richard and I had no such ties beyond the family. To soften the hurt I spent a lot of time in Helena's company talking about Australia. Sometimes we walked through oak groves to Tipperary town and back or we sat in the dappled sunlight of the walled kitchen garden with a pot of tea peeling potatoes for Mrs Ryan.

"Nicholas, promise me you will not compromise your safety by moving unguarded through native forests. I'm told the native black people of Australia are cannibals, and there are more tales of heathen Chinese who infest the lives of native tribesman and equip them with foul poisons and pagan spells."

"Helena, whoever informed you so?"

Helena stared at me forthrightly, "Kathleen. Our housemaid. She has two cousins there, in Van Diemen's Land."

"But I'm not going to Van Diemen's Land, Helena. We are bound for the new colony of Victoria. The reports from London deem it the most prosperous colony in the Queen's Empire. There will be Christian colonists and miners of all descriptions, strengthening company in numbers, and it is a lawful and orderly outpost of Great Britain. I believe from my reading that Melbourne, the main town of Port Phillip Bay, is larger than Tipperary. I do not believe there is anything to fear, but I will be mindful of your cautions, dear Helena."

Later we talked of kangaroos.

"And what of the huge hopping rats? The animals called kangaroos. Do you think them dangerous, Nicholas? May they transmit disease?"

"Not that I know of, Helena. But I'm reliably informed they are tractable and when treated kindly and well-nourished with honey they make excellent draught animals. Horses are in short supply in the colonies. I believe that many of the best families use kangaroos to draw their buggies. I'm told their hopping gait aids digestion."

Helena roared with laughter. We embraced. "I shall miss you Nicholas," she lisped.

I simply grinned. I had not yet attained the wiles of flattering sophistry. Helena arrived much later in Australia. She was unhappy and she disappeared. But that is a tale for another day.'

2

To Australia

Nicholas and brother Richard relish the excitement of Liverpool and her ships, seamen and touts. A Moroccan dwarf helps them find a ship offering free passage for Richard's service as a surgeon. John travels on another ship. Nicholas discovers seasickness, pretty girls, revolvers and loses a friend from the rigging of the ship in the great circle route in the Southern Ocean from South America.

Nicholas Sadleir continued.

'John departed first on a coach to Dublin and via a steamer across the Irish Sea to the English port of Liverpool. He arranged for a passage on the SS Great Britain for Australia and returned to Uncle Nicholas in Dublin before he joined the ship at Holyhead (he may have spent some time farewelling his sweetheart in Sligo again, but he never spoke of it). He wrote to Richard and me warning of the dangers of Liverpool.

No. 5 Great Denmark Street,
Dublin
1st August 1852

Dear Richard and Nicholas,
I have secured a single passage on the SS Great Britain in a second-class cabin for 30 guineas. It leaves for Hobson's Bay in the colony of Victoria from Holyhead on the 22nd inst. It is delayed because of necessary refurbishment for the Australian run. I will join her two or three days before then. I fear I have paid too much, but the ship is said to be luxurious, newly fitted, steam powered and should ensure a comfortable passage.
The steam packet from Dublin was overcrowded, and there was much sign of misery. It seems we are not the only people fleeing

Ireland. Thieves were abundant on the Dublin docks and they multiplied in number when we reached Liverpool. Have a care for your baggage. Ensure that it is stored below by the purser who will issue you a written ticket of ownership, or keep it with you and not let it out of your sight.

Employ the same prudence at debarkation in Liverpool. Porters abound. Choose one who shows a badge of licence, direct him with your baggage to Mrs Laidley's lodging house and walk with him to her premises. If Mrs Laidley cannot provide lodging, there are several lodging houses of equal quality nearby. Pay no more than a penny to the porter. I lodged with Mrs Laidley for one shilling a day. Victualling was satisfactory and I think this was a fair bargain.

There is an array of ships seemingly weighing anchor for Australia every day. It is important you choose wisely. Some may be unsafe or riven with disease. I recommend you inspect the ship and peruse the bill of fare before you commit to buying a passage. Further, I recommend the Tapscott Line of packet ships. Mr Tapscott dealt with me in a straightforward and friendly manner – as you know I chose to travel on the Great Britain, but that was not because the Tapscott packages were unsatisfactory but rather I preferred the novelty and luxury of this new and wonderful steamer with its steel hull.

There is favourable news for you, Richard. The Tapscott Line offers free cabins for surgeons. You may be able to work your passage.

A final word. You may need to tarry at least a week for a suitable passage. Low company abounds on the docks and in public houses. Avoid it. There are parts of Liverpool that are cultured and civilised far from the docks and there is a fine library. I walked there from the docks each day in all weathers to preserve my physique and morality.

I look forward to greeting you on your arrival. Uncle Nicholas has written to Redmond Barry, an attorney-at-law in Melbourne to arrange Nicholas's apprenticeship in Mr Barry's Chambers. When I arrive in Melbourne, I shall call at his chambers and leave particulars of my lodgings. When you debark, seek his chambers and thus find me.

I propose, as well, an additional method as a means of assurance. You will surely choose a ship days before you embark for Australia. Mail packets leave daily on ships for the colonies, and thus a missive from you telling me of your ship is likely to arrive before your ship graces Hobson's Bay. Address the letter to John Sadleir Esq, c/- Mr Redmond Barry, Attorney at Law, Melbourne.

Your affectionate brother.

John Sadleir

Richard and I departed Brookville House on a summer morning in 1852. The sun shone. The garden reeked with new flowers and I did not want to go. The spoken farewells were muted. We had done them together in many ways in the days before as the household fussed about the clothing and equipment we would need on the voyage, the suitability of luggage, the medicines and tonics we should carry with us, and our libraries for diversion on the voyage.

We descended from the steps to the driveway at the front of the house to Liam, the coachman, waiting at the buggy. Mater and Pater remained on the stairs. Helena stood behind them. We had all embraced in the hall before we descended the stairs. We exchanged not a word and tried to look composed.

The buggy had our baggage and trunks. We climbed in, raised our faces to the group on the stairs of the house, raised our hands and Liam drove us away quickly to the coaching station in Tipperary town. We never saw our parents or Brookville House again.

Liverpool was my first taste of England. It seemed viler than Dublin. There were crowds of people of all sorts. Families from Ireland sat on their luggage in groups on the docks in bewilderment waiting for agents to help them on their way to accommodation and passages on ships paid for in Dublin without existence in Liverpool. Rival agents harried them extolling the virtues of their ships. Children begged in rags, easy women plied their trade in alleys and doorways. There was filth, stench, and flies. Drunks reeled and vomited. Carters parted the crowds with cargoes and victuals for ships. There were even drovers with flocks of sheep. Everywhere traffic flowed and swayed.

We saw female convicts under guard being loaded on a ship bound for Western Australia. Most of them were Irish women. I remembered one as a cousin of one of our tenants and wondered what she had done.

Richard guided me, "Keep your hands in your pockets Nicholas. Keep your money safe. Look ahead. Appear as if you know where you are bound."

I interrupted. 'You were a bit green Holas?'

'We both were Blue. We'd had easy lives. I was 17, but Richard, although he was older, had been wandering about the scholarly chambers of Dublin as a student of Trinity or in philosophical conversations with friends. We were educated, but all that meant on the Liverpool docks is that we could read things that others couldn't.

'Not a lot of help with pickpockets Holas?'

'Yes Blue. You have the right impression.'

He continued. 'We walked with our porter and his handcart carrying our luggage to Mrs Laidlaw's lodging house as John had advised. Mrs Laidlaw welcomed us (she was from Waterford) but she had no room. "Sure but I think Mrs Murphy next door has room. Wait here, gentlemen. Polly, dart to Mrs Murphy and see if she has accommodation for two fine Tipperary gentlemen." And she turned to us. "Mind, I do not recommend the quality of her table, but you will be getting worse on ships, and do not pay her a penny more than 11 pence."

Mrs Murphy had room. "But, sure I have only one large bed. You have the look of brothers. I doubt sharing a bed will be a burden. It is a large bed and soft, and the room is airy."

Richard looked at me with a friendly frown when I offered Mrs Murphy 1/6 daily for the bed, and board for two people, and accepted her counter stipulation that we pay for two nights in advance – three shillings. "You have commercial acumen Nicholas; perhaps the colonies will suit you." The porter struggled up the stairs with our travelling chests to a low ceilinged attic with a small window overlooking a small yard with rows of cabbages. There was a bed and nothing else. We promised Mrs Murphy to return for our supper at seven and explored the docks.

The masts and rigging of the ships mesmerised me. What seemed a thousand bare tree trunks soared to a bright sky forming a spiky forest. Some of the spars crossing them were wreathed with grey sails, others were bare but they all had lines soaring aloft from tethers on the edge of the deck ending in neat coils. I ached staring at them. I longed to know their function.

Seamen working aloft called to us. "Paddy, will you join us for the glorious view. You may see the Isle of Man from here and almost to Belfast." Or, "Join us, comrades; the beautiful women of the Philippines and India await you. Come aboard. Preview the ship."

I moved to do so and Richard took my arm and muttered. "Have a care Nicholas. I agree we are looking for a ship, but as passengers. We are not bred to be seamen. I do not recollect you previously harking for a life at sea. I think we should find a ship of our choosing tomorrow. That is preferable to having a crew of a ship choose us."

I was suitably admonished, so we waved to the crew and sauntered on. But my enthusiasm seemed to infect Richard. We stared at and chatted incessantly about the elegance, plainness and difference of the craft we saw crowded together at this terminus. We evaluated clippers, barks, barquentines, gaff rigged schooners, ketches and brigs and we saw ocean-going paddle steamers. We were full of interest and excitement when we returned to supper at our lodging house. We had forgotten Tipperary.

I cannot remember our supper. It was obviously unremarkable, but I recall the bed. It was lumpy and Richard was in it with me, both in nightshirts. We could not sleep so we talked about our plans for finding a ship. We agreed that Richard would initiate interviews with agents as a surgeon seeking a cabin and dining room service for two, in exchange for his service as a surgeon to passengers and crew on the voyage. As a contingency, we proposed to offer a single steerage fare to compensate the shipping line for my additional presence according to our commercial instincts. It took me hours to persuade Richard to adopt this plan. He was hesitant. He had not firmly decided to work at sea as a surgeon caring for the crew and passengers. He doubted his capabilities.

"But Richard, my esteemed brother, you tended the sick for months in the poorhouse at Tipperary as part of your early years of study. You are experienced in the treatment of common ailments."

"In the poorhouse it was more like helping people to die in relative comfort. I had little medicine. And there wasn't enough food to help the sick recover."

"If you had medicine, would you feel able to take the post?"

"Yes, but I would need instruments and dosing measures as well. I have limited equipment in my sea chest. I'm not firmly resolved to practice medicine in Victoria. There may be other preferments."

"Richard, I believe there are shipping laws stipulating the medicines and surgical instruments a ship with passengers must have. If the shipping company must provide these, and does, that, surely, must set aside your objections."

"But are you certain, Nicholas?"

"No. But we may ask at the Board of Trade offices tomorrow. We passed the building on our walk this afternoon. Are we agreed on that?"

"Yes. That would please me. Now try to go to sleep, Nicholas. Your pestering has me exhausted."

I rested contently on my back. At 29, Richard was 12 years older. I had no right to believe that I could influence him. But I had.

Mrs Murphy got us out of the house at eight in the morning soon after a breakfast of oatmeal and tea in a long, flagged-floored kitchen. "Sure I'll not have you idling here when there is a house to clean. Dinner is at noon. Do not be late."

The Board of Trade office was a warren of clerks sitting, or standing at desks, surrounded by messengers who streamed from the street and bustled for attention. It was dirty, noisy and ill-smelling although well-lit with tall windows. It seemed too busy to be well maintained. We stood as gentlemen, holding our hats, waiting to be interviewed. No one came to us at first, and then a small Moorish looking urchin wearing a messenger's bag came to Richard with a swaying gait. "Do you want something done, sir?" he said.

"Yes," Richard said, "I would like to examine a copy of the proclamation regulating the service of medical and surgical care on vessels carrying passengers."

The urchin wasn't a child. He was an adult dwarf with a brown face sparkling with intelligence. "I know the one, Sir. Give me a shilling. Return in an hour and I will have it for you to peruse."

He kept his word. He waited on us by removing a group of messengers lazing at a reading desk by poking them with a short stick he carried and raging at them. "Be off you idle carrion. These are gentlemen of a noble purpose."

And the dwarf could read. He opened a bound copy of a book of proclamations he carried, rifled through the pages and pointed with a ringed finger to a page entitled ***Proclamation on the Passengers' Act*** and ran his finger down lengthy lists entitled ***Medicines, Instruments & C*** and ***Medical Comforts.*** The lists tallied supplies needed for 50 *statute adults* ranging from 2 ounces of Acid Nitric to two and a half gallons of brandy. There were descriptions of amputating instruments, sets of splints, lancets, enemas, a bleeding porringer, and tins of meat – there were more than 100 entries. Richard made notes. He handed the volume back to our able guide. "Thank you my good man. Can you direct us to the shipping offices of the Tapscott Line?"

"I surely can. It is far from here, you will need a cab. I am known to the people in the offices of the Tapscott line. Perhaps I can be of further service there?" He seemed to be appraising us. "Are you seeking posts as surgeons or simply a safe passage, gentlemen?"

"You seem to offer a range of opportunities my man," I said. "May we know your name and your credentials?"

"I am Isaac of Morocco. I was educated in languages, literature, and Euclid by Swiss Jesuits at Gibraltar and I came to Liverpool as a secretary to a Genoan wine merchant who employed me from the monastery. Sadly, he is no more. I live now as a general factotum – a go-between or negotiant for gentlemen such as you and for foreign gentlemen for whom I work as an interpreter and translator of contracts and bills of lading. If I may know of your plans and desires further, I may tell you how I may be of assistance." He looked at us directly, from one to the other.

I glanced at Richard, who nodded.

"We are Richard," I nodded to Richard, "and Nicholas," I touched my chest, "Sadleir, late of Brookville House, Tipperary, Ireland. We too are educated," I said, "at Midleton College in County Cork, Ireland, and for Richard, at Trinity College, Dublin. We are seeking passage to Port Phillip in the new Australian colony of Victoria. My brother Richard is a surgeon. He may be amenable to service as a physician and surgeon on the voyage for a consideration. We have good reports and observations of the Tapscott line. We believe if the next ship of that line bound for the Port Phillip district is in need of a surgeon we may be rewarded with considerations eliminating payment for our passage and providing superior dining and accommodation."

He paused and looked at his feet. I noticed his black-haired, balding pate carried oil. It shone and reeked of bay leaves. He raised his head and addressed us directly.

"I hope to be able to help you, gentlemen. If you agree, I will go directly to the Tapscott Line's offices. It is better that I go alone to negotiate favourable terms and conditions. If I arrange an appropriate consideration, I will ask you for one guinea each as a fee. If I can find you nothing to your advantage, I will seek no payment. May I know the name of your lodging house? I will call there after supper this evening to inform you of progress. May I have your permission to proceed?"

We nodded. Richard stooped to shake his hand. He noted the address of Mrs Murphy's lodging house in a pocket book he drew from inside the long blue greatcoat he wore. His pencil hovered above the page. "Mr Sadleir I shall need something of the breadth of your medical acumen, not all the details, just sufficient to demonstrate that you have received adequate instruction and education." Richard told him of his studies at Trinity College Dublin and of his experience in hospitals in Dublin and Cork.

"To this evening, gentlemen." Isaac of Morocco pocketed his book, reached to shake our hands and turned with a rolling gait to engage a hackney cab in the cobbled street outside. It was only as he walked away that we remembered he was a dwarf.'

'Was it good business, Holas?'

'Most satisfactory, Blue. Isaac earned his guineas. Richard signed as a surgeon for cabin passengers and crew – there was another aged surgeon who had already been engaged and his terms were changed to direct him in the care of steerage passengers only. It all went on very well and we waited only four days for the ship to sail.

Isaac was a remarkable fellow, a foreigner, a dwarf and a Jew making his way in one of the most wicked cities in the world. I think, from the commercial experience I developed much later in Australia, he got a fee from the shipping line for finding Richard as well as the guineas he got from us.'

'Remarkable you found him.'

'Perhaps he found us.'

Richard dispatched this letter to the family at Brookville House.'

The Packet Ship Cambridge
Liverpool Docks
8th of August 1852

My Dear Parents,

You will be pleased to hear that Nicholas and I have found a safe, clean and well-found ship leaving for Port Phillip Bay in the colony of Victoria the day after tomorrow. She is the Cambridge, a fast, reliable and safe packet previously sailing regularly with passengers and mail to New York in North America. There are good reports about her speed and safety. As always, good progress to Australia will depend on fair winds but we hope to arrive in the colony of Victoria within 90 days.

Unlike most passengers, Nicholas and I have the luxury of a private cabin and we will be served prepared meals in the officers' mess. We were fortunate to obtain the services of a Moroccan factotum who treated with the shipping company to employ me as a surgeon on the voyage in exchange for a cabin for Nicholas and me and prepared foods regularly provided. Most passengers are not so well nurtured. The ship provides scant daily rations and thoughtful and prudent families supplement these with small luxuries they carry aboard, but they must carry their daily rations to a communal ship's galley to have the food prepared for them. The family carries the food back to its quarters to eat it.

We will be among passengers living in first-class accommodation in the private cabins. There are second-class cabins too, but about 200 voyagers will live in steerage below decks in long dormitories with bunks in tiers at each side and seating at long tables between the bunks. The aspect is gloomy and airless and passengers housed here have limited time allowed for exercise on deck (we have no such restriction); however all parts of the ship are clean, sweet smelling and seemingly free of vermin. I have inspected stores of foodstuffs and found them wholesome and we carry adequate supplies of medicines.

I pray that our passage is free of cholera (there is recent news of several outbreaks on ships to Australia). The preparation and maintenance of the ship may assure our arrival in good health and my duty is to assist all aboard in the maintenance of their health and humour. I am optimistic.

I seek your prayers for our safe arrival.

With all my love and affection to you and all others in the household. Nicholas sends his love and affection too.

Your son,

Richard Sadleir.

Nicholas described the departure.

'It took all day for passengers to join the ship. Most of us were young men but there were families with women and children as well. All except first-class passengers stayed in their quarters as the crew prepared the ship for sea. We were privileged to be allowed to watch from the afterdeck. It was mid-afternoon and the weather was fair. A steam tug towed the Cambridge from the dock into the Irish Sea. I stood at the stern with some others, behind the helmsman. Richard was below attending to a woman with a sick infant.

The crew was about 30. A captain and two mates (I forget their names) six apprentice seamen, and about 15 seamen. And there was a carpenter, a sailmaker and various stewards and cooks (I was never sure of them, I met only a few of them – they were mostly below decks and worked during the day). The seamen divided into two watches, called the port and starboard watches, each with their own mate to command them. They served in alternating patterns around the clock, in four-hour watches.

One watch tended the ship as the tug towed the Cambridge up the Mersey River. The other stayed below. All seemed orderly. The mate called clear precise commands and barefooted seamen leapt into the rigging to release gaskets from sails, hauled yards to their sailing positions, tightened halyards, and adjusted sails to create comforting full-breasted shapes from sagging flopping canvas. More sails went on. All was orderly and measured. The crew enjoyed it. There was an optimism about it. The mate stood watchful and composed and soon we were making way under sail. We gained on the tug. The towing hawser slackened. Two seamen cast it off. We headed south in fair winds around Holyhead into the Irish Sea.

We headed south in fair winds around Holyhead into the Irish Sea.

I spoke to Richard in our cabin. He had been treating a woman with an ill thrifty baby and had been able to supplement the woman's diet. He was optimistic and seemed pleased to be underway.

"Although I feel burdened by the responsibility for the health of about 150 souls, Nicholas, I find it difficult to explain my light spirits. Perhaps it is because uncertainty is gone. We are committed now. I look forward to it."

"It is certainly an orderly ship, Richard. I can say that for it, although its motion disconcerts me. I think I will remain here. Please excuse me from dinner."

And I encountered my first serious bout of mal-de-mer, an affliction that was to dog me for most of my life. I rested in our cabin, retching, and without food, for 36 hours. I imagined death. Richard mocked me, but I was too ill to heed it. But it passed and I can remember my first good meal at sea – a lunch of salt pork, sauerkraut and potatoes with dried figs for dessert. By then we were somewhere in the mid-Atlantic starting on the "great circle route" skirting South America and leading us into the great southern latitudes where the ship could make good speed east to Australia.

I settled into leisure. I read. I studied the routines of seamanship. Some of the apprentice seaman shared their knowledge of elementary navigation and cordage and I formed other friendships. The captain permitted ladies from all classes to walk on the lower after-deck. As a first-class passenger, I had the run of the ship, except for the working areas on the upper deck, so I was able to meet some of the fairer sex and engage in gentle themes. In steerage young ladies' sleeping berths were of four, six or eight bunks in tiers per compartment and placed on one side of the ship.'

'How did you discover that, Holas?

I got no answer.

'Was her name Kathleen?'

Silence.

'Forgive my impertinence, Holas. I'm not imputing any misbehaviour but rather reminding you of your youngest daughter Kathleen who was kind to me when I was a small boy and visited Brookville House as a young woman.'

'I think you may have a cultivated turn of phrase Blue – more like your counterpart at Wilcannia trying to talk himself in or out of the Wilcannia lockup after a spree. But you have diverted me from my mid-Atlantic thoughts. Tell me about Kathleen's impressions of Brookville House.'

'I can, Holas. She wrote them in letters to Mary, her sister.'

Tipperary town is the filthiest place you ever saw, and I could never write a song about it[1].

'What did she mean by a song?'

'Oh, of course you don't know it. It was a Great War marching song called *It's a Long Way to Tipperary.*' I sang a few chorus lines:

It's a long way to Tipperary,
It's a long way to go.

It's a long way to Tipperary
To the sweetest girl I know!
Goodbye, Piccadilly,
Farewell, Leicester Square!
It's a long long way to Tipperary,
But my heart's right there.

He said nothing.

'I think Kathleen had a good time in Tipperary. She stayed at Brookville House and felt at home. Here is some more of what she wrote:'

They say nothing has been all that good for the last 50 years, but the estate is out of debt, and Marshal has worked awfully hard. He is a great kid, and we had good yarns together. He wants to get out of Ireland, but does not seem to think it right to leave his mother just as she is comfortable, after she has been through so much. She is 70 and tells him she won't last long, just wait until she is gone, but she hasn't a grey hair in her head and looks about 50.

'Thank you. Good news.'

Maude drove me into the town with her for shopping, and we met the local butcher, who was there in Pater's day, and he nearly jumped over the horse's neck when he knew who I was, and the ponies got a terrible fright.

'Who do you think he may have been?'

'Isn't it terrible? I can't recall or imagine him. But I'm enjoying Kathleen's words. I enjoy news of my children. Obviously I missed parts of their lives.'

Nicholas fell silent. Then he resumed. 'We crossed the equator and there was lots of fun. One of the mates dressed as Neptune and a couple of apprentice seamen went overboard to commemorate their first crossing, but there was nothing cruel about it, the captain saw to that, and the passengers were just spectators – not participants.

In many ways, the worst part of the journey arrived when we encountered the doldrums. No wind. Not a breath for days and days and I was hotter than I had ever been. It was unbearable below decks so nearly everyone gathered on deck to find shade beneath the slack sails. The only good thing was some target practice we had with new Colt revolvers; and another ship bound for Liverpool from Sydney in

New South Wales took off its way near us. It was a steamer. The captain sent a boat to us and offered to carry mail home. Out came the writing desks. After an hour of flurried writing, our captain sent a boat across to the S.S. Chusan with a bag of mail from us aboard the Cambridge to be delivered to our homes in England, Ireland, Wales and Scotland (and a few to Poland, France, Spain and Germany as well), and we received a packet of mail from the Chusan for the Australian colonies. I penned a letter to the family at home. Richard was busy with passengers suffering heat prostration so I passed the family his news as well:'

The ship Cambridge
In the mid-Atlantic (somewhere near the equator)
1st September 1852

My Dear Family,

I trust you got Richard's letter from Liverpool. He is busy attending to passengers' ills so I have been assigned as scribe to give you news from us both.

I expect everyone is surprised to receive a letter from us so soon. We are becalmed in the doldrums (about half way to Port Phillip Bay I believe) and a steamer bound for Liverpool has agreed to carry mail from us to our homes in Great Britain. We are incommoded. We have had no wind for three days and we sit in a sweltering heat with nothing to do but wait for any sign of a breeze. Not so our mail deliverer. She, carrying coal for her locomotion, has no need of wind. She carries few passengers but she has a small cargo of wool and some gold I think. I met the captain of the Chusan who boarded us from a small boat but he seemed secretive and gave only vague details of the cargo he carried. However, he was a charming fellow and he lunched with us in the officers' mess after a long meeting with our captain exchanging advice and information..

Apart from the doldrums' galling delay (our captain tells us it is more normal than not and the only cure for it is patience – fair winds will arrive when they are ready) the voyage has been without serious incident. Richard has lost only one patient, a young man from County Cork who sadly succumbed to consumption about two weeks out from Liverpool. He was buried at sea. The captain led a short service by reading parts of the Roman Catholic requiem mass. It was sad and moving.

I estimate about half the ship's passengers are young men bound for the goldfields in the new colony of Victoria and there are many countrymen from Kildare, Clare, Dublin and Wexford. Strangely, Richard and I are the only passengers from Tipperary. As for the rest: there is a sprinkling of Frenchmen, Germans, Spaniards and Danish. There is even a Polish count, but most of the people are from England, Wales and Scotland.

Families aboard do not seem quite so adventurously fortune seeking. For the most part, they comprise tradesmen seeking to begin small affairs in the new colonies with the assistance and responsibility of their wives and families. They listen to the talk of gold wisely, but most of their talk is about the establishment of premises and workshops and housing for their families.

Richard has a small deck cabin assigned to him as a surgery and he uses it daily for consultations with passengers and crew each morning between 10 and noon. Apart from emergencies, he is at leisure for the rest of the time. We spend our time in conversation with young men like us planning and imagining life in the new colonies.

We are looking forward to our landfall in a few weeks. We are well and in good spirits.

With love and best wishes to everyone at Brookville House from Richard and me.

Your son, brother and true friend,
Nicholas Sadleir.

'You didn't talk about the target practice, Holas?'

'Yes. Well. My parents did not approve of my interest in firearms. I thought it best not to mention our target practice in the doldrums in my letter home. We had an American armourer on board with a cargo of pistols he intended to sell in Melbourne. He was a talented man of commerce. His name was Aaron Goldstein and he persuaded the captain to have a boat tow a small empty barrel about 50 yards distant from the ship to use as a target. He made a small charge of a shilling for a loaded revolver – that was five shots, and he promised to deduct payments made for shooting if customers bought a pistol for ten guineas. Aaron said the normal price was 15 guineas, but because the pistols would have been used, he was prepared to let us have them at a discount of five guineas, and less the money we spent trying out the pistols.'

'What kind of pistols were they.'

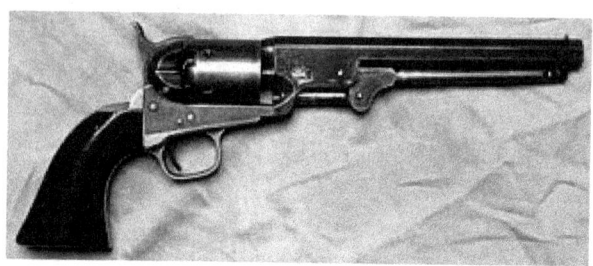

1851 Navy Colt

'They were Colt revolvers – handsome, modern American weapons. He had 1851 Navy revolvers – .36 calibre. One loaded them by pressing a ball atop a charge of black powder into five of six chambers in the revolving cylinder with a levered ramrod and then by covering the ball with wax or lard to keep moisture out and to stop accidental firing. One cylinder stayed empty for safety. Then one fitted a percussion cap to each of the five loaded cylinders. Aaron had six pistols on deck and he kept himself busy reloading them for his willing customers. He recruited me to collect the money and to have one shooter at a time to prevent accidents – he couldn't reload and conduct commerce at the same time. It took him five minutes to reload each pistol when he began, but he got faster with practice.

It was immensely popular. It ran from breakfast until mid-afternoon. We forgot about the heat. Some adventurous ladies took part. One, a mother of three from Surrey, a cabinetmaker's wife, struck the barrel with each of her five balls. But eventually the captain put a stop to it.'

'Why?'

'The Polish Count shot himself in the foot and the pistol ball damaged the deck. Brother Richard dressed the wound and the ship's carpenter came immediately to repair the deck. It was a good amusement while it lasted. The Navy Colt was a beautiful weapon.

That Count was a wonderful example of useless aristocracy who should be banned at all costs from attempting anything practical. He put an end to good sport. He succeeded on the goldfields and became a member of the Melbourne establishment, but he had a limp for life.'

'Did Aaron sell many pistols?'

'It's hard to say how many, Blue, but I know a variety of people on board bought them. And some of the purchases were secret. I got one and I used it for many years until I replaced it with a more modern Colt with cartridges. I had a pair of ancient duelling pistols in my sea chest. Aaron took them for ten guineas and sold the Colt to me for seven because I helped him with the sport on deck when I collected the money for him – you could say I got a new revolver at a profit. We took more than £20 in fees to shoot at the barrel. I think Aaron profited on the trip. The captain accepted a pistol as a gift and Aaron set up in Melbourne as an armourer and as an agent for Colt's Patent Firearms Manufacturing Company. He was there for years.'

'What happened to your Navy Colt?'

'I think I made a present of it to one of the Barkindji shepherds at Albemarle for kangaroo hunting when he grew too old to use a spear. His name was old Johnno. I can't remember his proper Barkindji name.'

'Any other vivid memories, Holas?'

'Well you try dying and telling stories a hundred years later, Bluey. Things do fade!'

He paused. Then he went on.

'We left the doldrums eventually and made progress to the east coast of South America. We called briefly at Rio de Janeiro for water and food but no passengers went ashore and we went south and west from there for a fast passage.

The Southern Ocean thrilled and terrified me. Gales howled and drove at the sails in a way that made me think the masts and rigging would be swept away, but all that happened was that the ship went faster and faster. We regularly logged 300 miles a day, and an apprentice seaman disappeared from the rigging one night as he went aloft to shorten sail. He was gone. The ship did not go about to search for him. Everyone on the watch that night knew he wouldn't be found. We didn't hear about it for several days. He was

John Watt from Yorkshire. We had formed a friendship. He secretly planned to jump ship when we got to Port Phillip Bay to go to the gold diggings with me as a partner. We had kept it secret, so I kept my grief secret too. And the crew didn't talk about him. They must have been ashamed about not turning back to look for him. I thought it apprised them of the danger each time they went aloft. I wanted to get off the ship as soon as I could.

---

[1] These quotations are from a series of letters Kathleen wrote to her sister Mary in South Australia, as she toured Europe and America in 1925.

3

Victoria and Gold

Nicholas and Richard leave their ship in nervousness and shock at the extravagance of the gold-rich colony but they settle in Melbourne and seek brother John. They meet Redmond Barry, a judge known to a Dublin uncle, who offers Nicholas an apprenticeship with an attorney-at-law. Nicholas declines. Richard starts medical practice in Melbourne. John joins the Victoria Police. Nicholas goes alone to the goldfields. He mines with partners successfully, holidays with relatives in Tasmania, and decides to explore pastoralism despite his aptitude in law.

Nicholas described his and Richard's landfall in Victoria.

'The Cambridge dropped anchor off Hobson's Bay on the 30th of October 1852 just after noon – 81 days out of Liverpool. It was hot with a drying north wind. Sand on the beach glistened and beyond that, there was a miserable looking grey stunted forest of shrubs with dry undergrowth. The buildings of Melbourne were just in view. The place looked dull but there was vigorous activity and we soon forgot the strangeness of the place. We shared the bay with more than 50 other ships, and lighters scurried from wharves to the ships carrying passengers and baggage ashore and loading empty ships with cargo.

Because we were in first class, Richard and I were among the first passengers to leave the ship. Seamen laid a flimsy gangplank from the Cambridge to a small steam lighter. It wobbled and we felt insecure as we inched downward to the open deck of the lighter. Our sea chests slid down the plank towards us and we caught and stowed them. More passengers arrived with their baggage. Just as we cast off, a whooping howl came from two of the Cambridge crew as they leapt from the side and landed nimbly among us. "It's gold for us," they shouted back to the ship. They had nothing with them and they were still barefooted with shoes hanging from laces around their necks.

They had deserted the Cambridge and left behind their pay for the passage. I thought they were stupid, and I must have looked at them reprovingly because Harry, who knew me, came to me and said, "We are not completely senseless Nicholas. The perils of life aloft in the Roaring 40s don't compare with the nuggets we can pick up here. Why should we stay? We are free men again."

I looked at them slowly. "Yes" I said, "but how will you live until you find the nuggets? You have nothing with you."

He grinned. "Look at this." And he pulled a cloth purse from his pocket and tipped out a handful of sovereigns. This is lighter and more useful than baggage."

He smiled at me again and turned his attention to the gangplank as we approached the Melbourne wharfs. He and his companion helped the crew of the lighter position the gangplank and they rushed up it to be first off. "Farewell, Brothers Sadleir. I hope you share in our certain good fortune." And the two disappeared beyond the crush of people, horses, drays and piling merchandise. Boards of timber, bales of clothing, casks of spirits, cases of candles and boxes of glassware flowed from lighters crowding the docks.

Richard and I felt strange with solid land beneath our feet for the first time for months. We dragged our sea chests to the shade beneath the wall of the customs house and sat on them waiting for our land legs. It was hot and we sweated, removing our coats and wiping our throats with our handkerchiefs. Later we removed our cravats and waistcoats. We had dressed formally for a grand arrival – it was rather overdone. The seasons were opposite – in Tipperary it may have been snowing.

"What do you hope to do, Nicholas?" Richard said.

"Perhaps I could pose the same question to you, Richard. I suppose we should look for temporary lodgings before we look for John. He should be here by now."

"I suppose the best plan is to find temporary lodgings somewhere near Collins Street. We think Redmond Barry has his Chambers there. Pater informed us of that. Nicholas, will you find a cab to take us to lodgings near Collins Street in the centre of Melbourne?"

He was my elder brother so I acquiesced. There was a long line of buggies and drays waiting for customers on a dusty strip about 100 yards from the docks and an equally long line of people queuing to hire them. A tout demanded a fee of a shilling to ensure my place in the line. I paid it. There was no choice – he had two assistants bustling people away from the line if they refused. Many left in disgust, but I saw several who relented and paid his fee as they assessed their choices. Even so, I had a long wait and it was late afternoon as we approached the centre of town in a shabby buggy drawn by a rangy bay mare more suited to a plough than to high stepping urban transport. Proper cabs were rare, the driver, Michael, said as he drew level with me, "We have some hansoms around Melbourne sir, but I counsel you to employ me. You may be after waiting for hours for one to arrive. I know a good lodging house, bed only for four shillings, and there are good cheap eating houses nearby."

Michael was a Belfast man. I hired him and we threaded our way tediously through a moving meadow of pedestrians and vehicles as we moved towards Richard sitting on the luggage. He'd been waiting in the crowd for two hours and was testy. I could see his red face. Quickly I led Michael to Richard and introduced him. "Richard, may I introduce Michael Gallagher late of Belfast. He knows Melbourne well and will conduct us to a well-run boarding house in the middle of the town. He knows Redmond Barry's chambers. He will give us directions so that we can go there in the morning to collect our mail and discover John." Richard calmed visibly – he was too well mannered to continue his pique. "I'm charmed to meet a fellow countryman, Michael." He lifted our trunks to the baggage tray and sat in the driver's seat beside Michael who pointed out buildings of interest along the way. He ignored me.

The dusty streets thronged with people. We saw what we guessed to be successful miners on street corners eating banknotes between slices of bread and offering glasses of brandy to all comers. There were gaudily dressed women with bands playing tunes to groups of miners who paid them for each tune by tossing sovereigns to them and the women scrabbled for them in the dirt. About half the people seemed the worse for drink and it was only late afternoon – about five o'clock. I thought we'd descended into Hades – albeit a cheerful one. There was no sign of piety.

About an hour from the Port of Melbourne Michael stopped the buggy in front of a humble cottage in Little Collins Street. He swung down from his seat and entered the front door without knocking. He emerged a minute later ahead of a stout woman wearing a full grey dress with long sleeves and a white apron. She looked about 50. Michael told us she was his sister-in-law. She was immediately businesslike. "Sure you are countrymen, but I must warn you that I charge four shillings nightly for a bed – in advance."

I turned to Michael and protested. "Michael surely this is excessive. We paid less than a shilling in Liverpool and that included full board."

"Brothers Sadleir, this is Melbourne. My sister-in-law is not trying to cheat you. She is an honest woman and is simply warning you about prices here. Everyone and everything is at the goldfields. Ethel has to pay five times the old going rate to find housemaids. I pay 10 times what I paid two months ago to get my old horse shod and the price of chaff has gone through the roof. The colony cannot grow enough hay for chaff. We import cut chaff in bags from South Australia. Last week there was even a consignment from the Cape."[1]

I relented. "We are from the same island Michael. We need to trust each other. I daresay we will both be in the colony for a long time and we will remember these circumstances and appraise them in the future. We will pay what you ask."

We unloaded. Mrs Gallagher showed us to our room, dirt-floored, white-washed with two beds and a table below a small window dividing them. It was so small we were forced to stack our chests atop each other to fit the space. We paid her, she gave us a key and we went for our first walk for pleasure on dry land in the southern hemisphere.

Richard spoke first. "It seems prosperity has its price Nicholas. We both need to find gainful employment soon. Our funds will not last."

"I agree Richard, but I crave fresh meat. Is that a steakhouse?" Richard put his arm around my shoulders and we charged into the eating-house together – another guinea for the first fresh meat we had eaten in months. We ate huge beefsteaks at a long table with teamsters, sailors, miners, builders and slaughtermen. There were languages from all over the world but we felt welcome and served ourselves from huge communal plates that female waiters in long skirts constantly replenished with beef, new potatoes and cabbage.

We found Redmond Barry's chambers easily the following morning. They were on the ground floor of a handsome, new three storey building faced in light orange stone with a large lobby leading off to several private suites of rooms. A board inside the lobby indicated his presence along with several other attorneys at law. He was not simply an attorney-at-law as we expected, but a judge. Nonetheless, he received us kindly. When we entered his lobby and asked his clerk if he could see us he followed his clerk immediately into the lobby, ushered us into his chambers and seated us facing him across a shiny mahogany desk free of papers but flanked by tall bookshelves filled with hidebound tomes.

He smiled. "Welcome to the colony, gentlemen. I have anticipated your arrival since I received a letter from an esteemed colleague, Nicholas Sadleir of Dublin. He is your uncle?"

We nodded.

"Which of you is his namesake, he mentions in his note that one of you may be seeking a post as an apprentice attorney, and is well capable of fulfilling that role?" He looked up.

I blushed and met his gaze. He smiled at me kindly. "I am, Sir," I said. "But I'm unsure of my opportunities and prospects."

"Are you seeking your fortune, Nicholas?" He paused when I blushed again. "I can see your youthful energy. You have time to appreciate the law as some of your juvenile urges dissipate. When you decide I shall be happy to interview you and make appropriate introductions. I judge the time is not right for you now, but I stand to be contradicted and we can proceed tomorrow if you wish."

"Thank you, Mr Barry. I shall think about what you said."

Mr Barry looked at the letter before him again. "I was expecting three of you. You are Richard and Nicholas. Where is John?"

Richard spoke. "John left on another ship at about the same time as we did – a steamer – a faster ship than ours. I imagined he would be here. We were expecting to find a letter from him here, Mr Barry. When we arrived, we asked your clerk if there was a letter or message from him. Your clerk found no letters or messages for us but there was a letter for John from us telling him of our ship from Liverpool.

"What is the name of John's ship?"

"The SS Great Britain."

"I can imagine your concern," and Mr Barry called to his clerk "Tompkins, could you send a messenger to the shipping exchange to discover whether the SS Great Britain has arrived? Gentlemen, please pardon my brusqueness, I have some urgent matters to attend to. May I suggest we continue this interview in two hours when we have news of the SS Great Britain? Perhaps you will join me for luncheon then?"

And so we walked the streets of Melbourne. It had a rawness about it. Few of the streets were paved and there were ruts where wagons and drays had been bogged. There were buildings in all stages of construction and they varied from ostentatious and grand to simple and humble. Many shops were under construction. We saw rows of tents in the distance and wondered who lived in them. It was the busiest town we had ever seen. People called to us and asked us where we were bound, but they were friendly. There was less formality than we were accustomed to. People seemed to wish each other well. Perhaps it was the prosperity.

We waited in the lobby of Redmond Barry's chambers. He emerged from his rooms after about 10 minutes with a note in his hand. "The SS Great Britain is not arrived. Do you know when she departed?"

"John informed us in a letter his ship was due to sail on the 22nd of August from Holyhead."

Mr Barry looked up. He calculated, "If the Great Britain left on time, she is 70 days out of Holyhead. I imagine her somewhere off the coast of Western Australia now. If she arrived in 10 days it would be a fast and satisfactory trip. It is too early to be anxious about your brother."

He beamed at us and spread his arms. "Now let us go to luncheon."

He took us to the Melbourne Club installed on the first floor of a nearby hotel. It was a short walk along Collins Street.

"This is a club for gentlemen, gentlemen. Perhaps when you are ready I may propose you for membership, but in the meantime the luncheon is excellent." The dining room was full. We heard accents of England, Ireland, Scotland and Wales and the nasal voices of the colonial born, but all the voices were lower and more restrained than at the eating-house we had visited the night before. Sturdy looking men almost furtively strived to appear genteel.

We had fish the waiter said was whiting, it was good, but it differed from the whiting we were used to from the coast of Ireland (yes it was King George Whiting, Blue, I expected you to interrupt) followed by mutton chops and a desert of plum pudding. There was even claret from Bordeaux. It was the best meal we'd had since we left Brookville house. (I did become a Melbourne Club member, but it was years after this, and in new premises at the other end of Collins Street; I never lived in Melbourne, nor did Redmond Barry propose me as a member, but I used the club for accommodation on my trips between Tasmania and New South Wales and Queensland – but that is another story, Blue).

We finished luncheon. On the steps of the club Redmond Barry expressed a wish that John would arrive safely soon, directed us to the shipping offices where we would have news of his ship's arrival, hoped we would bring him to his chambers when he arrived, and gave Richard brief advice about beginning a surgical practice in Melbourne. "Indeed there are opportunities almost everywhere, Richard. But I believe Melbourne will become a stable centre of civilisation as time goes on and you would be well advised to establish yourself here."

Richard heeded him. He wasted no time in the period we waited for the Great Britain. Within two days he had rented premises in the developing suburb of South Yarra to begin a medical practice, a short buggy ride from the centre of Melbourne. It was a small four-roomed single-storey building facing a narrow unpaved dusty street. There was no garden, just dust. He arranged one of the front rooms as a surgery and offered me a bed in a back room. I was still undecided about my future, I looked to the coaches leaving with jolly fellows for the goldfields, but I waited for John. If he was interested in searching for gold, I wanted to work with him. In those days I thought him strong and practical. I relied on his guidance and advice as a younger brother did. He had been my strength at Midleton in my first days there.

After three days at Michael Gallagher's sister-in-law's lodging house we moved to Richard's new premises away from the city with our sea chests on the back of a dray. The house was partly furnished. We travelled to the city to buy additional goods. Most of the available furniture came from England. Much of it was shabby and expensive, but we were able to find a skilled German cabinetmaker in a small shop off Little Collins Street whose work was good and whose prices were fair. He used handsome colonial timbers from Tasmania and New South Wales. Richard commissioned a large desk and two glass-fronted cabinets from a Tasmanian hardwood called Blackwood, one for medicines and one for instruments, for his surgery. It was a good affair. Fifteen guineas, and to be delivered to South Yarra in a fortnight.

Richard was assured. He beamed. "Nicholas, I have made good progress. You seem to have emerging business acumen and I thank you for it. Will you help me compose an announcement about the establishment of my medical practice for insertion in the newspapers?"

I walked to the shipping offices in Melbourne each day waiting for news of the SS Great Britain and so I was able to call on the offices of *The Argus* to look at copies of the newspaper and to see what form an advertisement should take. After several attempts, Richard agreed to pay a guinea for *The Argus* to publish this in two consecutive editions:

MR. Richard SADLEIR respectfully intimates to the inhabitants of Melbourne and its vicinity that he has commenced practice as Physician, Surgeon and Accoucheur, and may be consulted at his residence, 27 Forster Street, South Yarra.

I talked to miners on my walks. There were wonderful tales of nuggets and riches, and I saw the gold coming to town from Ballarat with an armed escort. A merchant from Devon sold mining equipment – picks, mattocks, shovels, cables, even wooden windlass rollers. He said, "Young man, Ballarat is by far the most popular field, and coaches leave from the coaching station four times daily, but I counsel you to consider Sandhurst. It is a newly developing field. I have recently sent a large consignment of mining equipment to my agent there. He has cleared it all and asks me for more."

And there was more: "the gold in Ballarat will never run out," "there are good reports from Ararat," ... "I'm just back from Avoca and look what I have." My mind was whirling. I was missing Tipperary.

On my eighth visit to the shipping office, on November 12, *SS Great Britain* was scribed in white chalk on a board labelled *Debarking*. As usual, the place was crowded and it took me minutes of struggling through the throng to get to the board. I ran to the wharf to join a crowd of people waiting for passengers and intrigued to see the new modern ship. The owners were said to have spared no expense in fitting it for fast passages on the Australian run from England. People chatted about it at the shipping office.

The ship was long, black, with five masts and one funnel. It was too far out to see much more, but lighters were leaving her and moving towards us with passengers and baggage. The wharf was a mass of humanity surrounded by baggage covering all available ground space. People moved with difficulty.

It took more than an hour for John to arrive on a lighter with 20 others. We hugged. John introduced me to his companion, a tall fair-headed man of military bearing. He was Jarrod Fox, formerly adjutant of the 75th regiment. John had met him aboard. I told John Richard had already leased premises and we had a bed for him, and I invited Jarrod Fox to join us as well. He declined, "Thank you, that is kind Nicholas, but I am expected at the Richmond Police Camp. I expect someone is here to meet me. I shall wait here. John, enjoy your family reunion. I hope we meet in a day or so."

I was puzzled. On the buggy we engaged to return to South Yarra, John talked of his journey but I was hardly able to listen. "It was an entirely unsatisfactory run, Nicholas. I do not believe the captain was a competent seaman or organiser at all. We were diverted to St Helena, we had to wait for coal for 10 days at the Cape and at one stage we appeared to be in peril of running aground on a reef. It is delightful to arrive, Nicholas. Is it always so dusty?"

I stared ahead. Eventually I blurted it out. "Do you plan an enterprise with Jarrod Fox, John?"

John turned to me. "Well yes, as a matter of fact I do, Nicholas, but why are you so anxious to know? I have just arrived. I am hardly settled."

I was embarrassed. "I imagined we would go gold-seeking together John. We dreamed about it at home before we left. How have you changed your inclinations?"

"I am sorry if I misled you, Nicholas. I was not firmly set on gold-mining but rather on the opportunities prosperity, coming from gold-mining, provided in a new colony. Several fellows on the Great Britain prevailed on me to join them as cadets in the newly forming Victoria Police. That is what I have resolved to do."

"And you are fixed on it."

"Entirely."

We finished the journey in silence. I suppose I was childlike in my disappointment, Blue. I matured a lot on that trip from the Melbourne docks. It was not my brothers and me. It was myself. They were good men with their own ambitions. I needed to get on with mine.

Two days later we three brothers parted. John set off with his possessions to the Richmond Police Camp. I left my sea chest with Richard, carried a bundle of clothes with my pistol, power, caps and balls in a soft bag with 20 sovereigns to the coaching station, and bought a ticket to take me on a Cobb and Co stage coach to Sandhurst.

We promised to write to each other, with our first letters going to Richard for him to be able to forward mail to us, and Richard said he would write home with our news so far.

Sandhurst became Bendigo later on, but that was after I relieved myself of gold-mining. I arrived in the second year after someone found gold there. The merchant from Melbourne who urged me to go there had an agent in what passed for the main street, which had about ten other buildings. He was in a tent united with a solid building made of timber slabs rammed into the earth and roofed with shingles with a dirt floor. The tent served as a shop front and the agent slept there as well. The timber building stored goods to sell and he kept the gold he bought there too until he could get it to a bank in Melbourne. It was empty. The agent's name was Frederick. He had a black woolly beard and wore a vermillion shirt and moleskin trousers. He was Polish but his English was passable. "You are new miner. If you want new pick, shovel et cetera, order it now. New stock coming next week. Write name here. Put what you want."

"Frederick, I'm new to mining. I'm not certain what I need to succeed. Can you counsel me?"

He stepped back and looked at me. "Yes. You new chum. Your clothes too good for mining. You have blankets? Bedding? Where you eat? Where you camp?"

He had humiliated me. My dander was up and he noticed. He held his palm out and pointed to a stool for me to sit on. I hesitated. Then I sat.

"Not to be flumbuxious young man. There is good man needs help. He has broken leg. His partner left and gone to Clunes. We put his leg in splint to mend but he cannot work, cut wood to cook, carry water and he no mine. He is good miner. Got plenty gold in Clunes. Spend it on woman in Melbourne. All gone. Now back for more. Good miner. He find gold. You walk back to coaching station, go straight to broken tree. Walk down gulley beside tree. His first tent you come to. He name Solly. He from London. Tell him Frederick send you."

Frederick turned his back and started to read a book. It was nearly midday. I rose and walked to where he told me to go. It was about a mile.

Solly sat with his back against a disused brandy cask on a stool with his leg in a splint supported by his bare heel resting on an overturned shovel in the doorway of his tent. Flies buzzed around his face. He was not pleased to see me.

"And who the hell are you?"

"Nicholas Sadleir."

"What are you after?"

"Gold."

"Well I haven't bloody well got any."

"Frederick sent me. He said you were a good and experienced miner. He intimated he thought you might be able to teach me something, and he told me he thought you were in need of assistance."

"What you think it farking looks like?"

I remained silent.

"Where are you from Nicholas Sadleir?"

"Ireland."

"Roman Catholic?"

"No."

"Makes no odds to me. I am of another ilk – Hebrew. Any problems?"

"No. What shall I call you Mr ————?"

"Leventhall, Solomon Leventhall, late of Soho, London, but call me Solly. All right Nicholas. Perhaps you are my guardian angel even if I don't believe in the bastards. What have you got with you?"

"Just some clothes."

"Got any money?"

"A little."

"All right. This is how we work. We both stump up 10 shillings a week for expenses. That's for tucker[2] and the few things we may need and we share the gold we get equally. Have you had a feed today?"

"No."

"Alright. Nor have I. The Cornish store is open. We need to stock up today as if we are starting from scratch. Can you read and write? Have you any writing material?"

"Yes." I rustled through my bag and found a pencil stump and writing paper.

"Ten pound of flour, five pound of spuds, two pound of sugar, a pound of tea, some onions if you can get them, two ounces of salt, five candles, a tin of jam and a flitch of bacon. Write that down. Here is my ten bob. Put it with yours and go down to the Cornish store beyond Frederick's premises. Open an account in the name of Solomon and Sadleir. Give Mrs Edyvean the pound on account and check the order against it. That should leave us a few bob in tick. You will need some proper working clothes. The clothes will be for you to buy, but buy our stores first and open our account before you start asking about clothes for you. Mrs Edyvean will size you up. If she believes you will become a regular customer, you will get your working clothes at a reasonable price. Tread warily. She can be a real bitch, but she has a heart of gold."

"Anything else?"

"Yair. Get the fire going before you go." He pointed to a wood heap of dead trees and to the fireplace a couple of yards from the entrance to the tent. It felt good to swing an axe. I came to the open hearth with an armful of wood. Solly tossed me a steel and flint and I found some tinder. "You know how to use these, Nicholas?"

It was wonderful to be able to nod. This was something I could do. Kathleen, the housemaid, had shown me how to do it as a small boy as she lit the morning fires at Brookville House. Solly leaned forward to watch me, grinning. He was small, balding, clean-shaven with side whiskers and twinkling dark brown eyes. Despite his obvious injury, his clothes – a grey flannel undershirt, blue cotton shirt and moleskin trousers – were clean and the tent was tidy. Somehow, I felt a weight lifting from me. I thought, *I could persist.* Solly seemed generous and intelligent. He wanted money from me, he came from the dreaded London Jewry, but something made me trust him.'

'Perhaps you had no choice, Cholas.'

'Maybe, but somehow one makes one's own good fortune, Blue.'

'How was Mrs Edyvean?'

'A handsome woman of medium build and middle age – in fact she was medium everything, except for her acuity which was extremely sharp. She wore a full skirt with a large white apron over a full greyish brown skirt topped with a sleeveless spotlessly white bodice. Everything about her was trim, neat and clean and she stood behind the counter with her hands on it as if she was mustering the contents of her closely stocked shelves to fall in obediently in line behind her. She knew who I was. "So you're the young gentleman who has come to help our Solly," she said as I walked in the door with my list. I don't know how she knew. Frederick must have told her he sent me to Solly's camp. "Welcome to Bendigo Gully. Thank you for your order. We do not run a charity here. I expect your count to be kept in credit with an accounting on the last day of every month. Mr Leventhall retains credit in the account he ran with his last partner

Simon Black who has cleared off to Clunes. Will you please ask Mr Leventhall what arrangements he prefers about the monies held in that account? I know he will be unable to attend this establishment for some weeks.

And now you will need some proper miner's clothes. You are well turned out. Indeed you seem a gentleman, but the vestments you bear are for more formal occasions. My word, you are a tall young fellow." And she produced a tape measure.

I carried my bundle of clothes and the groceries in a sugar bag back to Solly's camp. He showed me where to air the bacon with the onions in a sugarbag in a tree nearby and to fill the box under his bed with the rest of the stores. He showed me how to mix flour and water, baking soda and salt for a damper we cooked in his camp oven and ate with jam and sweet black tea.

"Mrs Edyvean was pleasant." I said.

"Good," said Solly. "She is a good judge. Had she not taken to you, she would have refused your custom and you would not be here now. She can be terribly fierce. Her prices are the best on this goldfield but woe betide a miner who lets his account fall into arrears, and she will not abide drunkenness. Beware, young gentleman Nicholas." And he smirked. "If you're going to be a miner it's best that you look like one. Duck into the tent and change your clothes."

We finished the first of many meals we had together on the Bendigo diggings and I cleaned our plates and the camp oven. Solly turned, dragged himself backwards with his hands along the dirt floor of the tent and indicated my bunk – two poles made of saplings supported by forked stumps driven into the earth and threaded into used flour bags to form a sagging bunk. There was a roll of blankets and a sugar bag stuffed with leaves for a pillow. "This is my last mate's bunk. It's yours now for as long as we last. Keep your clothes in your bag under the bed. If you have anything valuable, bury it, and don't let anyone see you do it. Most of the miners are good coves, but we get a rotten apple from time to time."

The bed was surprisingly comfortable. I rose before dawn, carried my clothes outside wearing my underclothes, dressed and dug a hole a yard to the east of our cooking hearth in which I buried the box containing my Colt pistol, balls, caps and black powder with 10 sovereigns I slipped beside them. I wrapped the box in a scrap of canvas I found behind the tent.

When it was light, and when Solly was awake, I made him a pair of crutches from forked saplings I cut with an axe from the thick eucalyptus forest less than 100 yards from the camp. He tried them. We padded them with rags, adjusted them, and by midday Solly was moving about in some comfort –able to attend to his ablutions with separate dignity in the forest away from the camp, and he directed me in preparing meals. I filled our water cask with buckets I carried with a shoulder yoke from a stream about 200 yards away. I gathered wood for the fire and cleaned our cooking and eating utensils. I washed our clothes. Solly watched. "You make a fine housekeeper, Nicholas. But what about doing some mining? Follow me."

The mineshaft was behind the tent about 20 yards away. It was hardly a shaft. Someone had started digging but there was only a square hole measuring six feet square and less than a yard deep. "This is where I broke me bloody leg. A goanna ran past and startled me. I stepped back with one foot in the pit and it just kept going down. The foot landed at an angle. I think I broke my shinbone and the one behind it near the ankle. I sort of hopped and dragged myself up to the broken tree. God knows how I did it, but a couple of blokes came by with a horse and dray and carted me down to the pub, dosed me with brandy and got Frederick to straighten my leg and put splints on it. They did it on top of the counter. There were three blokes holding me down."

"How is it feeling now?"

"A bloody sight better than then."

Solly stroked his whiskers and rubbed his beak-like nose. He had settled on a pile of red-brown earth beside the beginnings of the shaft. He played with the handle of a pick. He looked at me and he looked into the shaft. "You want gold, Nicholas. There is probably some for us down there."

"How deep will we need to go?"

"Probably no more than 20 feet. If we get no gold at that level, we will try another shaft."

"I will dig Solly, and willingly, but will you expand my understanding of why you chose to dig here?

"There is a little science, Master Nicholas, and a fair bit of hit and miss. As you walked down the creek this morning to get water you may have noticed signs of people scratching around in the sides of the creek. Some of the first blokes in the field got alluvial gold there just by digging it out and using a water pan to separate the gold from the dirt. In sinking the shaft here I am banking on a seam of gold running beneath it and leading to the creek.

"Will you demonstrate the science to me, Solly?"

"When I can walk Nicholas – probably after we finish this shaft."

And so we started to dig – or at least I did. I sank another two yards in the first day. On the second day Solly and I set up a windlass so I could fill buckets of earth and Solly could winch them up, standing on one leg and leaning on a crutch.

I had never laboured. I got fitter and my hands got harder. Solly made me piss on them. We worked as long as there was daylight. I went into town once a week to buy rations and to see if there was any mail at the post office. Solly got a letter from a cousin in London begging for money in the first week that I was on the Bendigo diggings but I had to wait more than 10 weeks to have mail relayed to me from Richard in South Yarra (and that was after I had written to him to tell him where I was).

I interrupted. "Teenager, youth, young gentleman, Holas – we've already had this dispute about what you called yourself, but how did you feel? I'd imagine this would have been a life you had never lived. Gentlemen didn't do what you were doing – even young gentleman. Obviously you adapted, but tell me about it."

"Blue, I shall have to take time to think about that, it's so long ago. I can remember aching all over, from the sustained physical exertion – I'd never felt like that before, but I didn't think about the change of my station in life compared with my life in Tipperary. I think it was mainly because everyone's life had changed – or at least the work we did had. There were medical men, lawyers, even clergyman working on the mines as common labourers like me, but in their manners and conversations and modes of address they still behaved according to the styles of their class. But in work and wealth, everyone behaved according to the money they had. Titled gentry cleaned the boots of tradesmen when they were down on their luck. Gold was everything and most people lived as their own men when they were mining for it. There were lots of masters but very few servants when the rush was on – it evened out again when the gold petered out – when I left the goldfields and tried pastoralism.

'Well how did the affair at Bendigo go?'

'Satisfactory. Solly's leg mended, but he always walked with a limp – one leg was shorter than the other and his ankle was stiff. We engaged with another partner, Jacques Villeneuve from Normandy, to help us move earth more quickly, we harvested some gold regularly (enough to ensure that we were not living on our cash and allowing us to invest in a cradle to separate gold dust and nuggets from the gravel we raised and to pay our licenses) and we kept at it. After earning better than average wages for about two years we got a good lode – more than £1000 worth each. Solly went back to London, Jacques went home to France and I decided to look at country in Tasmania. Of all of Australia, it had a climate most resembling the one I knew in Tipperary.

I spent Christmas with Sadleir relatives there –William Sadleir and his wife Grace, who had moved to the Launceston district from Tipperary five years before I arrived. They kept a small farm and they were happy to let me stay (I educated some horses for them and helped with ploughing while I appraised farms and estates for sale). Tasmania grew fine wool and most crops, it had a mild climate and there were fine estates that reminded me of those of Tipperary. I saw the Quamby estate then. It would have been ideal for me but it wasn't for sale and I didn't have the capital for it (little did I know that I would manage it for 13 years and that 11 of my 15 children would live there). I looked at other estates and they were all beyond my reach. I had funds to buy farms but not enough for an estate and I wanted a large spread, so I banked the money and went back to the goldfields to gain some more.'

'Back to Bendigo, Holas?'

'Yes at first. I had friends and colleagues there and I would have been welcomed in partnerships but the opportunities for alluvial mining were fading. Large companies drove deep shafts with expensive machinery. I wanted to keep to the system of low investment I knew. Not much to spend on equipment – some acumen – and some luck and hard work. So I removed to Ararat for two years and spent the last year at Stawell. I formed a partnership with five fellows at Ararat and we found good money shallow mining in shifts – two-man teams, one man in the shaft mining and another up top on the windlass and cradle, the other team resting and sleeping and one man cooking, shopping, carting wood and water, selling gold, keeping accounts and keeping the camp and tent tidy. We made wages, we exchanged jobs regularly and we trusted each other, but after two years the claims petered out, Chinese miners were taking some of the good ground, and we decamped to Stawell together.

We made one good strike at Stawell at the beginning and we banked £1000 each. But that was the last gold for us. After a year we parted. We were good friends and it was sad to disband but we lived a ruthless life, Blue. We never stopped for amusements. None of us married or had sweethearts. We just wanted gold and knew the only sure way to get it was to keep working. Tom Ellison, the oldest and the most experienced partner said at the end. "This may be the best thing to happen. We've decided to get out while we have savings. We can get on with another life. Who would want to spend the rest of their days gouging at a red earth wall and floor hoping to see some flecks of yellow without seeing the sky for days."

'It was agreeable we ended like that. We saw some terrible fights. Some partners killed each other dividing assets. We just sold or left what we didn't want and shared the cash that remained.'

'How did you agree on a method, Holas?'

'In the beginning, we had two things to talk about – how we divided the work we did and how much cash we needed to contribute weekly for tucker, licences and other expenses. It went without saying we divided the gold equally. One pound a week each was straightforward and plain enough to keep the mine running and us in tucker – we didn't buy grog or tobacco, if anyone wanted it they paid for it themselves. The "one up and one down" shift was pretty plain, two gangs working 12-hour shifts, but since there were five of us, there was a spare man to conduct the camp and keep the books. I was the only one who could read and write easily so I proposed a system of bookkeeping – just a simple ledger of income and outgoings. We agreed to put in one pound a week for consumables and when we got gold, we weighed it together, sold it, declared a dividend, and divided the money equally. When it was somebody's turn to run the camp, if they couldn't read and write, they would ask me to make the entries for them. I always did as they instructed and made sure they understood that I did it. I suppose in a way I started to teach them to read. It was the good records we kept and the way we accepted our share of the work that held us together and avoided arguments. I learnt a lot about the importance of good records in that last partnership. Sticking to it over the years kept things running well by and large, Blue.

I had seen the end arriving at Stawell. All the Victorian fields were yielding less gold every year, and there were more and more miners.[3] We were all edgy because we had been so long without gold.

Six weeks before we broke up, I wrote to John Phelps at Cannally station on the Murrumbidgee River in New South Wales. He was the distant cousin of my mother's from Limerick who had excited us with his letter to Tipperary telling us of his hundreds of thousands of sheep. I asked him about opportunities. He responded briefly:

Cannally station
Via Balranald
3rd July 1857

Dear Nicholas,
I am in receipt your letter dated 1 June 1858. I shall be pleased to discuss opportunities for you in pastoralism. I expect to be at Tarcoola station on the Darling River in September. Please call on me there.
Yours in good faith,
John. L. Phelps.
And that was the end of gold-mining for me.'

'And you still had the Navy Colt and the 10 guineas, Holas.'

'I did. Sharp of you to remember, Blue. We used the pistol at Ararat. Hordes of people came to the district week on week, about half of them were Chinese; small things went missing from around the camp and we were worried about losing the gold we were getting before we could sell it, so one Saturday morning we stopped work for about half an hour and I arranged target practice. We showed those watching we could hit a stringy bark tree at 30 paces. I had a holster made by a local saddler and the partner running the camp wore the loaded pistol in the holster at his belt. It was good insurance. We never discharged the pistol against anyone and we missed nothing from the camp.'

'You had profits from mining. Why did you not return home to invest there?'

'Unfavourable politics, Blue. The news from home became worse with each letter. Pater said this in his last to his sons in Victoria when I was in Stawell (we circulated the letters, first Richard, then John and me last):'

Brookville House
Tipperary
5th November 1856
My Dear Sons,

Thank you for your news and good reports of your progress. Your mother and I are elated by your various successes and we hope for further triumphs.

We are proud that your sister Helena has obtained a post at Kilkenny Castle tutoring young women and girls of the Butler family in the fine arts. She is looking forward to the position and left for Kilkenny to set up residence two days ago. The Duke of Ormond has a fine reputation and we have high hopes for Helena.

Winter approaches. Our summer harvests were good and many of our tenants and their cottiers had excellent potato crops. Your brother James continues to flourish in the management of the estates. Your mother and sisters are well. They continue to interest themselves in charities and good Christian works in the county.

The political outlook remains sombre. The Repeal Association, an insidious movement, still attracts mass support for a separate Irish parliament. I fear it may prevail with disastrous consequences for orderly government and commerce within the British Empire.

However there is worse news for the family. Cousin John Sadleir, a member of Parliament, was accused of malfeasance in the affairs of the Bank of Tipperary and has killed himself. Several weeks ago there was a run on the bank as depositors withdrew their funds and the bank is now closed and declared insolvent. James Sadleir, his brother, is implicated too and he has decamped to France. There is much local discontent. Many solid local families have lost most of their capital reserves.

The family name has been damaged, perhaps irreparably. Good local people of all religious and political persuasions now avoid us because our name is Sadleir. It is most unfortunate. James and John are distant relatives and Papists, and we have had no personal interest in the affairs of the Tipperary Bank – but no matter – the Sadleir good name, renowned for generations of service to the Church of Ireland and to the maintenance of mannerly law and order in County Tipperary is now associated with financial scandal.

For your sake, my sons, I am pleased you are not here. The shame is hard to bear. I hope it will pass, but for the moment times are unpleasant.

I am sorry to burden you with sour tidings, but it is likely the news of the scandal will spread to the colonies (it is all over London) and you need to be informed.

With all our love and blessings,
Your father,
James Sadleir.

'The news circulated in The Melbourne Argus, but we were able to faithfully answer to anyone who asked, that John and James Sadleir were distant relatives and that our family had no interest in banking. Few people enquired. Tipperary and the House of Commons in London were far away. As far as I know, it did no one called Sadleir in Victoria any damage.'

[1] The Cape of Good Hope in Africa

[2] Tucker means food in Australia.

[3] The Gold Field Commission reported that for the three years 1852, 1853 and 1854 the population of miners on the diggings grew from 35,000 to 100,000 and their average earnings fell from £420 to

£82. The total value of gold produced fell from nearly £15 million to less than £9 million.

4

The Riverina

Determined to join Limerick cousin John Phelps with his thousands of sheep, Nicholas again resists his brothers' wish to make him a Melbourne lawyer. He assists at brother John's marriage and leaves for Swan Hill on the Murray River. He meets Sam Curtis, a ticket-of-leave convict and a paddle steamer captain, and makes a friend for life. He takes a steamer to Wentworth and walks 100 miles to Tarcoola station with an Aboriginal guide to meet John Phelps. Nicholas impresses John with his skill with horses and John and Nicholas Chadwick start to show Nicholas the art and commerce of arid lands pastoralism. They share dreams and realities.

'How did you start off for the Darling River, Holas?'

'I went by stagecoach from Stawell to Melbourne to stay with Richard after we terminated the mining partnership.

Richard's affair was expanding. He had combined with other investors to establish an accouchement hospital with ten beds in South Yarra. He was still living in his old premises but he hoped to move to the hospital to live to supervise its good conduct. John had been promoted within the Victoria police and had moved to the new goldfields at Beechworth. I had not seen him since I'd been on the goldfields, but since his arrival he had been stationed at the Richmond police camp near Melbourne, at Ballarat, back to Melbourne as a sub inspector, and now to Beechworth.

Richard told me John's sweetheart, Isabella Crofton, had arrived from Ireland and was living with her brother near St Kilda. We had roles in John's wedding and he stayed with Richard and me at Richard's house in South Yarra the day before the wedding. We talked far into the night. I was pleased that John and Richard encouraged my plans to see what John Phelps had to offer about life in pastoralism for me. John said, "I have been worried about your life as a miner, Nicholas. I have seen too much misery, tragedy and felony on the goldfields and I feared ill might befall you. I'm pleased I was mistaken, and I'm equally pleased that you have decided to leave the life. But have a care. I'm told the West Darling region of New South Wales is wild and poorly governed. The blacks may be dangerous."

Richard was circumspect as well. "Like John, Nicholas, I'm pleased you have decided to leave gold-mining, but I urge you to reconsider the opportunity for you to become an attorney-at-law. Life in Melbourne is agreeable, profitable and safe. Why not stay here? You have money to invest in real estate, and property values increase each year. Remember, Redmond Barry left open his invitation to assist you in finding an apprenticeship."

I had not been harangued so directly for years. Life in the simplicity of a mining shaft suddenly seemed attractive again. I had reverted to somebody's little brother, but I curbed myself, Blue. "Thank you both. I have been lonely and it is kind of you to show me your filial concern. I still think about a legal life from time to time. I am but 23 years old. So far I prefer life away from towns, but I know I may return to Melbourne should a pastoral life not suit me." And I changed the subject and talked about the wedding.

I don't remember much about the wedding. It was at Christchurch, St Kilda. Richard was best man and I was a groomsman. Isabella looked splendid in her wedding dress and her brother, Harmon gave her away. There was a modest wedding feast at a nearby hotel and some formal speeches. One of the bridesmaids was pretty, but she chattered, and I was pleased when the feast ended. I gave the bride and groom a set of silver serving dishes imported from Germany. I bought them from a miner at Stawell who was down on his luck. It was good affair, I intended to sell them in Melbourne but John and Isabella seemed better recipients.

After the wedding I deposited £1600 in the Union Bank and enjoyed the luxuries of Melbourne for a week more – bath houses, barbers, some fine dining, a couple of stage shows, but I soon tired of it. I said goodbye to Richard, gathered my travelling chest, which had rested at Richard's house for all the years I was on the goldfields, filled it with some new clothes and books and loaded it onto a stagecoach bound for Sandhurst in the Bendigo goldfields. From Sandhurst, I planned to join a coach going north to the Swan Hill settlement on the River Murray and thence to a paddle steamer going along the Murray and Darling rivers. I carried 50 sovereigns in a money belt and the Navy Colt in a holster at my belt hidden beneath my coat.

Sandhurst had expanded to become a modern town. There were paved streets and bridges across creeks and rivers. Many of the old landmarks and shanties had gone but I was able to find Mrs Edyvean's establishment. It was expanded into a large, high-ceilinged, shiny-floored emporium with numerous shopkeeping staff and a range of luxury goods in glass cases as well as ordinary grocery items and clothing, but Mrs Edyvean was there and she remembered me. "Nicholas it must be more than three years, or is it four? My, you have developed into a fine young man. Has no Sheila or Colleen invited you to join her yet?"

I blushed: "Not yet, Mrs Edyvean."

"What became of Solly and Jacques? I always thought you three an ill-assorted trio, but you were all good men and I was pleased to see you get a good lode of gold and go off to other lives. Not that I wanted to lose your custom, mind – it so often happens that good fortune takes the best people away from these districts."

"I never heard from Jacques, but I got one letter from Solly. He married when he returned to London and I believe he is now a tea merchant and moneylender."

"And you went to Tasmania, Nicholas?"

"Yes I did, Mrs Edyvean. I looked at some farms but I decided I didn't have enough capital, so in the last few years I've been at Ararat and Stawell after more gold. I got a little more. Now I'm on my way to the Darling River to see what fortune awaits me in sheep, cattle and horses there."

"Well I'm glad you're not trying again here, Nicholas. This town is nearly a city and that is good for my family, but many of the diggers have fallen on hard times. Most of the shallow gold is gone and the large companies have taken over. If you came back you would probably be working for wages in a deep shaft for a company,

Nicholas. Thank you for calling, young man, and please take the time to make someone's daughter happy. I have things to do, and I expect you do too. Good luck and farewell." And she bustled off.

The next day's stage on the coach to Swan Hill was long and slow. There were only seven other passengers. Two Chinamen, "fee go to make garden swanill," (they rode on top) a police constable who was to join a station there (he knew my brother John) and a blacksmith who was looking for business along the Murray River, with his small new wife, they were on their honeymoon, they had been married in Geelong the week before – she was neat and pretty and her beauty embarrassed me – I can remember staring and then looking away as she caught my eye). There were also two boatmen who had responded to newspaper advertisements seeking their services in the riverboat trade. We were away from the main coach routes connecting the gold-mining towns with Melbourne. There were fewer horse-changing stations and so much of our progress came at a walking place to save the horses, and the road was little more than a track – rutted – and with more than its share of difficult creek crossings.'

We arrived at Swan Hill at 10.30 in the evening. It was cold. A frost was settling across the river. The creaking coach stopped outside the hotel with snorting, drooping horses. The coach driver said, "You'se'll find a good hot feed waiting for you. The cook here allus waits up for the coach. Ee's Chinese. One of the best cooks 'long the river." On hearing this, our two Chinese passengers rushed through the dining room and into the kitchen behind it. The sing-song chattering started, but it stopped when the publican, Adam Barton welcomed us all to a late supper – or those that remained of us. The two boatmen left the coach and walked towards the wharves across from the hotel beside the river. They hoped for a bunk and a feed on a riverboat. I promised to join them after supper to see if there was a prospect of a boat getting me to Wentworth, or even better, turning up the Darling when it got to Wentworth.

So the police officer, the blacksmith, his wife and I shared a delicious supper: beef consommé, delicately poached river fish, lamb cutlets with potatoes, carrots and peas and a desert of apple pie and cream. It was better than anything I had had in Melbourne – but perhaps it was the hunger. It had been a long time since lunch at the coaching station at Kerang.

It was nearly midnight. There was a frosty moon. I shivered and pulled my coat about me as I walked to the wharf. Both paddle steamers at the wharf showed lights.

A later river steamer – the PS Ruby.

They were side-wheel paddle steamers moored to the dock and facing each other – it appeared one was bound upstream for Echuca and the other steaming downriver and probably to the South Australian river port of Goolwa. It towed a barge loaded with wool lashed to its starboard side. The other steamer had no barge.

I waited at the dock beside the steamer facing upriver and called out, "Is there anyone there? Permission to come aboard?"

The door sprang open with a glare and a fog of brandy fumes. A grey-curly headed, stout, barefoot 50 year-old stumbled out, tripped and fell on the deck, looked up with one eye and tried to speak. "Whach ya want?" He raised himself, stretched, grasped an upright pillar with a raised hand and leaned on the pillar with his head against his arm.

I stood on the dock. "I'm looking for passage downriver. When do you leave?"

"Well don't just stand there. It's cold. Cum'n av a drink." He levered himself away from the post and raised his arms in the air. We have many things to discuss, the route we plan, the choice of music for the voyage, your culinary preferences and certain pecuniary arrangements."

I followed him as we ducked beneath a doorway and entered a room whose sole piece of furniture was a table about 10 feet long and with shelves for seating built into the wall. This must have been the mess room for the crew and passengers. It was white painted. A small stove about halfway along the opposite side of the table heated the room. Three others sat at the end of the table furthest from the door and ignored me. They held playing cards. Two had come with me to Swan Hill on the stagecoach. One of them raised a finger and nodded to me before he turned back to the other two. Grey Curlyhead indicated a seat, a glass and a bottle, "Siddown, we're in the middle of a hand. It's for big money. We won't be a minute."

Curlyhead lost. He threw his cards in front of the other three. "I'm giving it away. Youse bastards are too good for me. I'm up for a drink with our gentleman caller – as it were." And he tossed his head and gestured his hand like a chorus girl on the goldfields. He was comical. He poured me half a tumbler of brandy, filled his own glass and raised it. "Your name, Sir, and whom should we drink to?"

"I am Nicholas Sadleir, and may I suggest we drink to arriving at McLeod's Crossing?"

He raised his glass aloft and drank deeply. "A good toast, but it's not gunna do you a bit of good, Nicholas Sadleir, we're goin' up river to Albury to pick up wool and turn back for Goolwa at the bottom of the river in South Australia – be away for weeks. But you're in luck. I reckon the boat next door's going to McLeod's Crossing the day after t'morra with a load of chaff and flour to go with the wool it has for Goolwa. And neither the flour nor the chaff is here yet. They're waitin' on a wagon from Wycheproof. So that'll probably suit ya. Call on 'em in the morning. The crew don't drink much and they're probably in bed ba now." He paused, filled his glass, hovered the bottle over mine (I had taken no more than a sip) and continued.

"You're in no hurry at all. Y'll need to learn to relax and if you don't learn, your nerves will anyway; riverboat travel has a wondr'us calmin' 'fect. How long since you been on a boat?"

"The last one was the one I came to the colony on, but it was a ship."

"How long ago?"

"About five years ago. Liverpool to Melbourne in 81 days."

"So you came as a free man then?"

"I'm sorry. You have the advantage of me. I'm afraid I missed your name."

"No you didn't, Nicholas Sadleir. I didn't give it. Are you in the way of having a brother who is a trap[1] at Beechworth?"

"How does my brother concern us?"

"I thought as much, both Irish, la-de-da Irish mind you, but still Irish and you look the same. Why would I be giving my moniker to the brother of an Irish trap?"

"Because the brother of the trap isn't a trap. He may be a free man, but he means you no harm. He has taken a glass with you. He wants to know your name simply to be cordial, and he needs friends along the river. He means to settle in this district if it lives up to its promise. Now, will you tell me what I may call you?"

"If you take that pistol from your holster and put it outside the door."

"How will I know that there is not somebody outside who will steal it? What if I gave it to you for safekeeping? I do not judge you to be a murderer or a thief."

"My name is Sam Curtis and you can call me that, Nicholas Sadleir. Keep your pistol where it is. I think you're a good man too."

It was the start of a long friendship, Blue. The card players had been watching us during our confrontation, perhaps expecting trouble, but when we reconciled, they went off to bed. Sam and I talked for hours and we finished the bottle of brandy. Sam told me

---

1. Police Officer

he was transported from England 15 years before because he was convicted for violence to the overseer of a large estate he worked on in Yorkshire. He told me the estate overseer had been interfering with his sister. Sam had assaulted him and continued to do so periodically until the overseer was crippled and Sam was arrested. Sam arrived in Botany Bay 18 months later and was assigned to work on a farm in the Hawkesbury district west of Sydney town.

"The owners were good to me Nicholas, I became one of the family, and when I got my ticket of leave, they gave me 10 guineas to try my luck on the goldfields at Bathurst. I got very little there and so I crossed the river into Victoria – to Beechworth. That's how I know your brother. He and I had a run in about a horse I'd bought. He thought I bought it from a band of horse thieves picking up horses the Omeo district. I hadn't, and eventually Inspector Sadleir believed I hadn't, but it took a long time to sort out. He apologised, but ever since my transportation I've found it hard to befriend policemen."

"How did you finish up here?"

"When I got my ticket of leave I encouraged my sister to join me here from Yorkshire. She came out about six years ago and joined a household in Geelong as a housemaid. She took up with the groom of the household next door and a couple of years ago they bought a small farm out of Echuca with a bit of help from me. I came up to stay with Jess, Bert, and their two small children on holiday from mining and the riverboat game was just starting up. I went on one of the first trips up to Albury and back. I loved it. The work is easy most of the time and the river is beautiful, and if you treat them right, the blacks are no trouble. And I have a base with Jess and Bert on the farm at Echuca if I want to have a spell on dry land and try my hand as an uncle again."

I left the Bluebird at 2:30 in the morning. I was not completely clear-headed, but I felt I'd made an important friend, and I had. I worked with Sam Curtis and riverboats for all of our lives on the Darling and Murray rivers.

Sam Curtis and the Bluebird had left by the time I emerged from the hotel in the morning. Thomas Martin was the skipper of the other boat. The two other crewmen were Tony and Bert. Thomas agreed to take me to McLeod's Crossing. He said he would reduce the fare from £3 to £1 if I agreed to help them load and unload the chaff and flour he was waiting for to take there. "There is no cabin accommodation, Nicholas, we sleep in swags on deck, but I think you'll be comfortable on deck if you nestle into the chaff bags. We have tarpaulins and plenty of blankets."

The Gundagai was about 120 feet long and narrow – probably less than 30 feet wide. It was flat-bottomed too and it should have been unstable but paddle wheels on either side levelled it with the weight of the engine and the boiler. The only structure was towards the stern and it served as a wheelhouse, cookhouse, storehouse and mess room. The engine and boiler sat amidships in the open beneath a flat roof with a small funnel, and there was room for a sizeable stack of red gum cut in four-foot lengths to feed the boiler. Forward was an open deck for cargo. Thomas had lashed the barge carrying bales of wool beside the boat as a labour-saving measure. "We save a man, Nicholas. If we tow a barge, we need to steer it. And we need to rig a wheel on top of the load so the helmsman can see and we have to set up pulleys to connect with the rudder. That's a bit of fuss. If we find a river wide enough to allow us to be twice as stout as we are with hulls side by side, we do it every time. It isn't pretty, but it's easier."

We loaded the bagged flour and chaff in separate stacks forward of the wheelhouse the following day from a wagon that had arrived from Wycheproof that morning and lashed them down. The wagon had been on the road four days. The horses looked exhausted. The carter held back five bags of chaff to feed his team of four horses on the way home. "I reckon they'll look forward to an easier trip back," he said, "We had a fair load, and it took the sting out of them. I'll give them a spell here for a day before we head home."

We left for McCloud's Crossing (it was called Wentworth later) at six o'clock the following morning. The journey was about 300 miles; Thomas thought it would take about six days. "It depends on how many snags[2] there are and how much we can travel at night. If there are too many we tie up at the bank for the night. The other thing that holds us up, Nicholas, is firewood for the boiler. There is one family that cuts wood for me regularly about a day down river and I

---

2. Snags were usually fallen dead trees. Occasionally shipping companies shared steamers specially equipped to remove snags. From time to time governments contributed.

can rely on them, but from then on it will depend on who is there and what they have cut. I think there are a couple of gangs working closer to McLeod's Crossing cutting firewood for sale, but am uncertain about them. If the worst comes to the worst we will have to cut wood ourselves, and that will hold us up. But that's the charm of the river; it's improving – more people are starting to think that cutting wood for riverboats is not a bad way of making a living."

Thomas was about 45, of medium height, balding and wiry. He was a partner in the boat and he had a wife and four children at Goolwa in South Australia. He was looking forward to getting home. The crewmen lived at McLeod's Crossing, Thomas was worried that they would leave his boat there and that he would not be able to recruit additional crew. He was sober and cautious but stoic. "If my crew leave me I'll just have to wait for more. Are you up for a trip to Goolwa and back Nicholas? It's beautiful at this time of the year."

He was right about the beauty of the trip. The engine chugged, but it settled into a soporific routine and we glided on gurgling brown water beneath and beside towering eucalypts. Some of the old trees had butts of more than ten feet in diameter. There were long riverine curves and wide horizons. And then there were thickets of lignum to confine the view to contrast the vistas. Wild duck swarmed the river, pelicans abounded, and on the first night, when we stopped because of fear of snags, Tony caught a Murray Cod – about two feet long. He cooked it for us in coals on the riverbank and we ate it with sweet black tea and damper. It was a delicious, delicate fish.

On the third day out, we stopped to greet a band of blackfellows who gave us 20 wild duck for a bag of flour. They waved at us from the bank and held up the duck. "Is this a common trade?" I asked Thomas. "I've not experienced it before," he said, "but it looks as if somebody has started it. They knew they wanted a bag of flour and they knew that we would like duck. This riverboat trade has only been going for two or three years. It seems the amenities are improving already. Perhaps we have the last boat through here to thank for it."

"How do they get the duck, Thomas? They have no fowling pieces."

"I have no idea. I've heard they swim beneath them and catch their legs, but I don't believe it. Perhaps you can make a study of it, Nicholas, and tell me when we meet next. You tell me you hope to be here for a while."

And I discovered the method soon enough – I saw it in detail on the way from Wentworth to Tarcoola, but by the time I remembered to tell Thomas when I saw him again four years later, he knew it. The river people (I knew the Barkindji best – they were the people of the Darling) got the duck by using two teams of people. They found a narrow piece of river. One team pulled a net across the river with at least five hidden people holding it on either bank. Another group hid on either side of the river in the direction the duck would come from. As the duck flew down the river the people hidden on the bank threw sticks above them and make them swoop towards the water. At the same time the people holding the net quickly stretched it and raised it. The duck flew into it, broke their necks, or were enmeshed. It worked well. Sometimes they got more than 50 at a time. I can remember reflecting on it. The Barkindji got less and less duck in the time that I was there. The riverboats tended to kill the method as they became more numerous. Progress had its price.

The rest of the trip was uneventful. We got to Wentworth inside six days. It impressed me. People still called it by its old name – McLeod's Crossing. It was my first New South Welsh town and it had a different colonial presence. It was certainly larger than Swan Hill, but that was not surprising. It seemed older and was becoming the administrative centre of this far-flung part of the colony. It became a major port too later on.[3]

But I delayed my assessment of the colony of New South Wales; there was cargo to unload and Thomas was anxious to make more miles downriver that day. We got the chaff and flour off soon enough. Wagons set off to unload it at stores in town, and I stood with my foot on the sea chest and waved my three companions off as they tooted and steamed off down river. They were still on the Murray River. The Darling River flowing from the north beckoned

---

3. For many years Sydney and Newcastle were the only ports in New South Wales to handle more cargo than Wentworth. A total of 420 boats steamed into Wentworth in 1890.

me. A horse and trap took me from the wharf with my chest to a lodging house for no charge. The lodging house employed the driver to find customers.

After a breakfast of sweet black tea, salt beef and damper I went to the police camp to ask for advice about the best way to Tarcoola. The sergeant in charge was Robert Mitchell. He was of medium height, barrel-chested, with a magnificent black curly beard and he wore the heavy navy blue uniform of the New South Wales border police. He was busy unloading stores from a packhorse. "Excuse me, sir. I'll attend to you in a minute. I need to get these dry goods stowed away and the ammunition locked up. I've just come in east from a tour. Been away for 10 days."

I waited. The camp was on the edge of town on flat ground about 500 yards distant from the junction of the Murray and Darling rivers – a neat array of six tents surrounding a wooden slab hut that served as a storehouse. Robert Mitchell carried boxes from the packhorse into it. I saw no sign of a prison or jail, but 30 yards away one blackfellow rested in the shade of a large tree, tethered to its trunk by a chain fastened around his ankle. He was asleep. A dirty grey rug covered him and the remains of a meal – chop bones on a pewter plate – and a large empty pannikin lay on the grey clay beside him. Two other blackfellows in blue police uniforms led three horses each to a yard behind the tents. I imagined they had just returned from the river where they watered the horses. The horses were wet. Perhaps they'd been swimming in the river.

The sergeant bolted and locked the door, turned to me shook my hand and said, "Robert Mitchell, Sergeant in charge. How may I be of service?" He led me into a tent, sat in the swivelling chair behind a bare desk of rough timber, indicated a chair across the desk to me and looked at me for an answer.

"Nicholas Sadleir. I'm bound for Tarcoola station. I'm seeking your counsel about the best means of travel – the road to take, the pacification of blackfellows along the route, and any general precautions I must take to ensure my safety. I have arranged to meet John Phelps of Cannally station there. We are distant relatives from Ireland. Mr Phelps has offered to interview me about pastoral opportunities in the Western region of New South Wales."

He stared at me thoughtfully. He was appraising me. He held himself with the appearance of an experienced military man. I felt confident I would get good advice. "You are from Ireland, Sadleir. How long have you been in the colonies? How have you spent your time? What experience do you have of these Western Districts?"

I told him my story, how I had come from Tipperary with two brothers, the goldfields I'd worked on in Victoria, and my decision to stop mining. I said I'd come to the Western District of New South Wales to learn pastoralism, I hoped to invest in it, but I knew very little about the district, except that it provided good pasturage for sheep, cattle and horses, and there were still opportunities to take up grazing leases.

"How did you fare on the gold diggings?"

"Moderately. I have some savings."

"Are you alone? Do you have supplies, equipment, arms, ammunition?"

"Yes I am alone Sergeant Mitchell, I have no transport or equipment, but I have the means to purchase them if they are available. I am armed."

"Young sir. This is a hard and dangerous part of the world. People die here of thirst, loneliness and despair. Only the strong and single-minded prevail. You have just come from the relative luxury of the Victorian goldfields – do you really think moderate success there qualifies you to succeed or even survive here?"

He irritated me. I felt I was being questioned and addressed like a schoolboy who had missed some important point in the discussion of an algebraic problem. "Sergeant, I fail to follow your line of questioning."

We glared at each other. Then I dropped my gaze, paused, looked up and smiled at him. "Thank you for your cautions, Sergeant Mitchell. I thought for a moment you resented my enquiry, but I am of a mind now to believe that you care for my welfare and that you reinforce my need for my own self-assurance and responsibility. Be assured that I've considered the hazards of this adventure concordant with the knowledge I have. You have assured me of my wisdom in consulting you."

"Yes. Well." It was his turn to be discommoded. He paused.

"Let us begin with the journey. Tarcoola is on the Darling, roughly 100 miles north of here. It is not easy to miss if you know the country, but as you do not, you will need a guide. I doubt that you will procure horses in this town at present. One cannot buy them for any money and I'm not in a position to let you borrow any of ours. We hope for another drive of horses along the river from the Albury district before Christmas. Last year a reliable dealer promised us a drive of about 200 head of mixed sorts at about this time. Everyone in the district is waiting."

He had been looking off to the side and addressing a point somewhere wide of my left shoulder. But now he looked at me directly. "That will mean walking. I can find you a blackfellow guide who is returning to that country after a period with me. The journey will take five or six days. The weather is warming up but it is not too dangerous to walk at this time of the year and you will be following the river. You will not need to carry water. What do you think?"

"I had expected to be able to purchase horses Sergeant Mitchell, one to ride and a pack animal to carry my luggage. It seems I have no choice but to resort to Shank's pony. I accept your offer gratefully. I expect I can store the major part of my luggage somewhere in the town securely and return for it later. Can you advise me on the stores and equipment I may need to carry myself, what I need to provide for the guide and when I may expect to depart for Tarcoola?"

"I suggest you travel light, Sadleir – use a shoulder bag. Carry about two pound of flour, a twist of tea, salt and a bit of sugar and a couple of blankets for sleeping. Make sure you have a flint and steel to light fires. You will sleep on the ground in the open air. You have a hat with a broad brim – you will need it. The blackfellow will feed you with meat and he will probably get you some fish and fowl. He will eat damper. You said you were armed?"

"Yes. I have modern American revolver with ball, powder and caps in my luggage."

"Good. Carry it at all times. Do not use it against blackfellows unless you see your life in peril. I am responsible for their protection and well-being as well as yours. Your guide is here and can leave tomorrow. Come here carrying the things I've suggested, at 7.30

tomorrow morning ready to leave immediately. Good morning, Sadleir. If you wish you may store the remainder of your luggage and equipment here. I shall be out for the rest of the day, but I shall leave a message with the constable on duty to expect you should you choose to leave it with us. I shall not be here tomorrow morning but the same man will be here to introduce you to your guide and see you on your way tomorrow. Good luck." We shook hands.

I walked to the town, purchased two blankets, a leather shoulder bag and bits of flour, tea and sugar, packed the shoulder bag with clothes from my sea chest and hired a buggy to transport the chest to the police camp. Constable Bates greeted me. "Sergeant Mitchell told me to expect you Mr Sadleir. Here is a receipt for your chest. I look forward to seeing you at half past seven tomorrow morning."

"Thank you Constable. I have no aboriginal languages. How may I communicate with my guide?"

He laughed. "Sammy will know where to take you. He is looking forward to it. He's going home to his country. I reckon you two will get on well. Use sign language. And Sammy has a few words of English. If you're lucky you might pick up a few of his words along the way. He is a reliable fellow, well known to us. You should be as right as pie." And he returned to the journal on his desk. He made the whole affair look like a routine thing.

Of all the new things I had started in Australia, this was the most disconcerting. I was to entrust my fortunes to a black heathen with whom I could not communicate and I was to spend about five days alone with him. And worse. I had not met him. This was not the journey I had imagined. I slept badly that night Blue.'

'What did you imagine about the journey to Tarcoola when you left Melbourne, Holas?'

'Oh, I thought a paddle steamer would take me from Wentworth directly to Tarcoola station – right to a wharf near the homestead. The fellows on the steamer from Swan Hill helped that dream to dissipate. They told me steamers had never been up the Darling. It was just that no one had tried, Blue. In the nearly 50 years I stayed in this country I can't imagine the Darling River without paddle steamers when they had water to move in. I suppose there is a start to everything and in 1857 navigation on the Darling River had not been tried. What timid souls we men can be sometimes!

My trepidation was worse the following morning. I rose early. I put on miner's clothes and a light coat for pockets. Breakfasted. Checked my baggage. Verified the good order of my pistol and its loads. Holstered it at my belt. Checked everything again, hoisted my bedroll, bag and billy[4] and walked to the police camp. I was immediately horrified when, seeing my approach, Constable Bates walked to the prison tree, roused the prisoner to a standing position, bent to unlock the chain from his ankle and stood with him as if at a mayoral reception awaiting me, the distinguished guest. The blackfellow wore no clothes or shoes, although he did drape the filthy grey blanket I'd first seen him sleeping in across his shoulders. He was as tall as me. As I approached he looked at the ground.

"Mr Sadleir, may I present Sammy your trusted guide to Tarcoola. Sammy," and he took Sammy's right hand and placed it in mine. "This Boss Sadleir. You take Tarcoola directly."

I protested. "Constable this man is a prisoner. I assume he is a felon. How may I trust him?"

"He's not a felon, Sadleir. We brought him in because of cattle spearing, but we found that somebody else was doing it. We planned to let him go yesterday but when you turned up looking for a guide we kept him here for you. He's a good man. We've used him for tracking and as a guide before."

"But won't he be vengeful to me because you have chained him to a tree wrongfully?"

"No. Won't make any difference to him."

"But surely he is a human being?"

"Yes but a lot tougher than we are. Trust me. You will be as good as gold. Look at him. He is keen to get going."

And Sammy was. He looked at me directly for the first time, turned away, waved forward to the trees lining the river with his right hand and set off north at a brisk walk. He carried three spears in his left hand and a blanket over his shoulders.

---

4. A billy is a small metal bucket or pail used to boil water for tea in a campfire.

I looked at Constable Bates. "Go," he said. "You want to get to Tarcoola don't you?"

And so I set off after Sammy. "I'm coming," I shouted. He stopped and waited for me, took my right hand in his left hand and we walked together like this for about half an hour. I remember looking at his hand. It was black but I could see the veins and the pale fingernails. It was an interesting interlay of black on white; me and him. I was pleased no one was watching. I thought about it long afterwards for many years, Blue. I think he sensed my fear and was trying to calm me.

On the first day we stopped at a sandbank by the Darling late in the afternoon. The river ran beside it and there was a waterhole shaded by two red gums. I can remember thinking how beautiful it was compared with the ruinous disorder or the goldfields. Sammy gathered some dry wood, dropped it between us and started to light a fire by rubbing a small piece of wood into a larger log beside a pile of tinder. I realised what he was trying to do and produced my flint and steel. I lit the tinder. He stopped rubbing the wood. Stood. Smiled. Waved at the smoking fire with approbation, pointed at me and the fire and held his palm down indicating that I should stay with the fire, picked up his spears and walked over a sandhill away from the river. I stayed with the fire and built it. I guessed he had gone hunting.

He was back before dark carrying a small kangaroo across his shoulders with one hand and his spears in the other. He was quite naked. He had left his blanket and a small knitted fibre bag with me. I hadn't noticed the bag before. Sammy dropped the kangaroo beside the fire. He put more wood on the fire and urged me to do the same. I went for more wood. It lay everywhere as dry branches from dead saplings. As I built the fire Sammy disembowelled the kangaroo with a small cut in its abdomen. He didn't skin the kangaroo. It was a strange method of butchery. The guts came out of a small hole Sammy made with a piece of sharp stone he carried in the bag. He carried the guts away from the fire and returned to seal up the hole he made in the kangaroo's belly with a sharp stick. He used it skilfully – like a six inch skewering pin. Then he threw the kangaroo – skin and all – into the fire. The fire died to coals. He turned the kangaroo and covered it with embers. The hair on the hide singed, the hide cracked and blackened and he left the kangaroo there, cooking. After more than an hour he pulled it from the fire with its tail, cut into the hide above the loin with the sharpened stone, peeled back the skin and dug out a piece of steaming meat to hand to me. He grinned, "Good tucker", he said. I blew on it to cool it and started to eat it with my hands. It certainly wasn't fine dining, Blue, but it was the best kangaroo I'd tasted. I'd had some cooked with lard in a camp oven on the goldfields, but this was better.

Later, as night fell, I made a billy of tea and a cake of damper on the hot coals. Sammy accepted both. He relished the sweet tea. We slept. Sammy seemed dangerously close to the fire when he settled into his blanket. I mimed a remonstrance. He grinned, pointed at the fire and hugged himself. I realised I was not in charge, he was cold and needed the fire. We both slept soundly in our own way.

We travelled amicably. Sammy proudly showed me features of his country. We dug for an echidna and roasted and ate it. We met a band of his kinsmen who were hunting duck and allowed us to join them in raising a net across the river to snare the duck when they dipped in flight to avoid the sticks members of the band threw above them. We delayed our walk by nearly a day to do this, but I didn't object. The people charmed me. They encouraged me in my netmanship, and we had a wonderful feast of duck. The naked women did the cooking as their children played with sticks. They did not disembowel the birds or pluck their feathers. They simply encased them in a bundle of wet clay they got from the riverbank and put the clay bundles in coals until the clay cracked open. Then the duck was cooked. The intestines shrivelled inside the carcass and the feathers and skin came away with the clay. The birds were delicious and the cooks delighted in my pleasure in eating them.

I had glimpses at home, Blue, but these were the first naked women I'd seen entirely. I wasn't incommoded. It was the way that they were. They were not seducing me, nor me them. It felt to me they were my mothers and housemaids. They treated me like a boy.

In return, I made sweet tea for the band. There were about 15 and they all tasted it with obvious joy. They nodded and beamed their thanks at me. Everyone except me slept close to little fires with native dogs to keep them warm. Early next morning Sammy and I left without ceremony.

On the last day, Sammy let me come with him looking for kangaroo. As he was stalking in long grass, I kept still and silent, but I shot a kangaroo dead at about 30 paces with the Navy Colt before Sammy could launch his spear. Sammy jumped at the sound of the shot, birds soared from trees and the other kangaroos scattered. Then Sammy saw the dead kangaroo, and he looked at me and grinned. He approached. I pulled him beside me, pointed to the gun indicating that he should watch it, cocked the pistol and fired a ball into a tree about 20 yards away. I led Sammy by the hand, showed him the hole the pistol ball had made in the tree and showed him the wound in the dead kangaroo's chest. Then I led him to watch me from beside me. I fired a ball at the tree and handed him the pistol standing behind him with my arms outside his with my chin on his shoulder to guide his aiming of the pistol.

He reeked of smoke and sweet sweat.

I moved his finger to the trigger, but I think we fired the pistol together. The second time I stood behind him without touching him and he successfully sent a ball into the tree. He handed me the pistol, beaming.

I reflected on what I'd done afterwards, Blue. I'm sure the police back at McLeod's Crossing would have considered it a criminal act to teach a blackfellow to use a firearm during that period of settlement. Perhaps it was, but I don't think anyone but Sammy and I knew. I never informed anyone. It was part of my way of expressing my gratitude. I gave him a knife as well. He was delighted.

Sammy delivered me to Tarcoola the following morning. He left as soon as we saw the homestead. He pointed, touched my arm and left. I never saw him again. He went along the river further north. I don't know what happened to him, but I saw members of the band I met who taught me duck hunting afterwards. Some worked for me. When I asked about him, they pretended not to understand me. I imagined a tragedy, but I learned nothing until years later.

Tarcoola homestead had a temporary air, Blue, rather like a military bivouac. Huge red gums gave it a backdrop to the river and I walked past horse yards confining five hacks beside open-sided sheds with bough roofs containing saddlery, drays, tools and a rudimentary forge. Like the sheds, the homestead was open-sided, or at least it appeared to be. As I drew closer I saw it to be a large shed sheltering a smaller building made of poles driven into the ground and plastered with clay. It had a sturdy bolted door with a padlock. Ordinary items of furniture surrounded the small mud-plastered building almost giving it the status of a cupboard – and that's how it appeared to be. I thought it was a storeroom used to secure food and other valuable items. A fireplace with a stone chimney leant on one wall and faced me as I approached. A fire smouldered in its grate. Breakfast appeared to be over and the cooking area was tidy. Cooking pans hung in the chimney and tin plates stood in stacks on a table fronting the fireplace.

The place had an orderly welcoming air. Ordinary living took place in the open air sheltered from sun and rain by the large bough roof. There was a table and chairs and several stretcher beds rather like the one I'd used on the goldfields at Bendigo. The entire floor was swept dirt. I suppose the total living area covered about half the size of a tennis court. I was able to imagine it as elegant and palatial. It was remarkable how the five days of rough camping had changed my perceptions of luxury.

I approached the house. A dog ran to me and licked my hand. There was no one else. I hung my bag on a nail driven into one of the corner posts, draped my bed roll over it and walked towards the river behind the homestead to see if anyone was at home.

A sturdy, athletic looking man carrying a yoke with a bucket of water at either end scrambled up the riverbank and walked towards me. We stopped and examined each other. He said, "I am Nicholas Chadwick. And where have you sprung from young man?"

I said, "I am Nicholas Sadleir. My mother in Ireland is a distant relative of Mr John Lecky Phelps. Mr Phelps invited me here to give me advice about affairs of livestock in this region. Is he here?"

"He left early this morning to inspect some ewes and lambs with a shepherd about 10 miles upriver. I expect him back by dinner time. Welcome, Nicholas. You look as if you could take a cup of tea."

I followed Nicholas Chadwick to the hearth in the homestead, he unloaded his buckets, enlivened the fire, boiled a kettle, found a teapot, china cups and saucers, made tea and served it with a plate of johnnycakes. It was the first tea I'd had from a china cup since Melbourne.

"Tell me of you, Nicholas. While you still have the speech of old Ireland, and I love to hear it for Kildare is my home, you don't have the air of somebody who is straight off the boat. How did you come here? Have you been in the colonies long? What have you been doing?"

So I told him my story. I talked about my brothers in Victoria, my times on the goldfields and of my journey from McLeod's Crossing with Sammy and the help I had received from members of the police camp. Mr Chadwick sat quietly during the telling. When I finished he nodded and smiled, refilled my cup and offered me another johnny cake. "And tell me, Nicholas Sadleir, what is your assessment of this country here?"

"I fear I am no expert, Mr Chadwick. So far I have seen some country bordering the river and I am not properly acquainted with the pastures away from it, although Sammy and I did cross dryer country when we left bends in the river on our way here. I am from Tipperary – luscious pasturage – some of the best in all Ireland with the fine cattle and horses and wonderful crops of grain and spuds. I expect it would take me much longer to judge the usefulness of the herbage in this region for the sustenance of sheep and cattle and horses, and so far I imagine it is unsuitable for cropping, unless one can devise a way of using waters from the rivers."

His pale brown eyes twinkled at me beneath bushy grey eyebrows. Elbow slumped on the table, he leaned towards me and asked, "And what about its suitability for camels?"

He took me aback, but I smiled and shared his mirth. "I have never seen a camel, Mr Chadwick, let alone judge myself capable of prescribing suitable pasturage for the species, but I appreciate your levity. I judge the best knowledge I bring from Ireland of use to the Australian colonies is that in the care and management of horses. I noticed the horses in the yard looked well and I saw no sign of hay, grain or chaff. In fact I saw no mangers or water troughs. How do you keep them so well?"

"A horse tailer looks after them. He is away watering a herd of six at a waterhole up river. He will be back with them presently; he will leave them in the yard and take the others that are there to water. It will be dinnertime by then. After dinner he'll saddle one and ride it with the others to good pasturage we reserve for horses about two miles downriver. You may have seen signs of horses on your way here with Sammy."

"Yes I did. I thought we must be getting close to a station when I saw the dung."

"This is a good season. The horse tailer (I suppose you're used to calling him an ostler, but here it's quite a different responsibility) knows his job and he'll keep the horses in good condition for no more than a few hours a day on good natural feed. When the feed is poorer we need to graze them longer, and sometimes at night as well. We don't use chaff or grain. It would be nice to have it, but carrying it on bullock drays from the farming country in Victoria or South Australia costs too much."

"But there is a scant supply of green grass, Mr Chadwick."

"More than you think, Nicholas. And horses eat some of the other things beside grasses as well. There are quite a few esculent bushes. Horses seem to prefer them. Do the horses of Tipperary look better than these?" And he twinkled again.

I remember avoiding the question Blue by asking another. The horses in the yard certainly looked well, but I had the pride of the Tipperary to uphold, I avoided conceding the equality of its horseflesh. "Where do you get your horses, Mr Chadwick? What do you use them for on the station, and how many do you need?"

"My goodness, Nicholas, you do ask questions of a searching nature. We use horses for the same purposes as we do in Tipperary – to get about, to take stores out to our shepherds, to travel down to McLeod's Crossing, for droving to markets or to other runs we have.' He scratched his head and looked at me. "I don't think anyone else has asked me that question, ever. Horses are just part of us. And we never seem to have enough.

As for the other part of your question, where do we get them? Goodness knows. We can buy none for ready money. We walked the lot we have here now from Balranald – from Cannally station, John

Phelps's home, with some sheep we brought up from there about a year ago. It was good easy trip – beside a river all the way, and it took about three weeks – 2,500 mixed ewes and hoggets and about 15 horses.

We could breed horses here. We plan to do it. There is a shortage. John has a good entire[5] for us at Cannally. We hope to get him up here in a couple of months."

"Mr Chadwick, I notice the horses are not shod. Why do you not shoe them?"

"We do not have the roads of old Ireland, Nicholas, to wear out our horses' feet. The horses of Sydney town wear shoes and I expect they do in Melbourne as well because of the hard roads. For all the time I've been in this country I've not seen the need for it. Sometimes we need to trim hooves, but even that is rare. This country seems to have the right balance of sand and gravel to wear the hooves enough to keep them in good shape. It is a blessing. We save a lot of time and money."

I nodded and thought on it. The new colonies certainly had new ways.

Mr Chadwick took my cup and washed it with his in hot water in a basin he poured from the kettle on the hearth, dried the cups and settled them neatly on the table. He tossed the remaining johnnycakes to a dog that guzzled them, brushed the crumbs to the dirt floor, straightened and announced it was time for a formal inspection of the station homestead. "I expect you will be staying for a day or so, Nicholas. I had better give you the lie of the land."

We walked to the river behind the homestead. Steep-sided, it flowed in a narrow gutter to a large pool shaded by majestic red gums. "We had a strong river last year. It was halfway up the banks, but since we've been here, even when the river is sluggish, this has proved to be a permanent waterhole and we get water for the homestead here. John has plans to install a pump here to water an orchard and a vegetable garden. He has a wonderful show at Cannally. He has a Chinaman there who grows almost anything you can think of – oranges, apricots, grapes, plums, cabbages, corn,

---

5. An uncastrated male horse used for breeding.

potatoes, carrots – the lot. A Mrs Phelps may even get flowers for the homestead if she ever happens to exist, but somehow I doubt whether we're ready for that level of decor here at Tarcoola – but fruit and vegetables from time to time would be grand."

We walked to the bough sheds and the horse yards. There was a small building on the corner of one of the sheds made of abutting posts plastered with mud and used as a store in the same way as the one at the homestead. Mr Chadwick unlocked it and showed me inside. It had spare stirrup irons, two new saddles, lengths of steel bands, small kegs of nails, hammerheads, saws, and new axe heads, spades, shovels, coils of rope, sheep shears, sharpening stones, tins of strychnine, jars of neat's-foot oil, and several tanned ox hides.

"How do the stores come here, Mr Chadwick?"

"We've had one bullock team with a dray up from McLeod's Crossing since John and I took up this place. That brought up the bulk of the stores you can see, but a lot of the smaller stuff came up on packhorses last time I went down to McLeod's Crossing about a year ago."

We stopped talking. A horseman approached wide of the river. His tall bay hack swayed through knee-high grass, saltbush and blue bush and moved towards us at the horse yards.

"That's John Phelps, the man you've come to see Nicholas. I'll go back to the homestead to see if the cook has got back with a load of firewood and help him unload that. Wait here for John and perhaps you and he can follow me down to the homestead when you're ready."

I could see John Phelps now as his horse moved up the path towards the horse yards. He wore a high crowned broad brimmed grey hat, a dark blue heavy coat over a light grey blue shirt, moleskin trousers and dark brown knee boots. He looked tall with a long face, long straight nose and a full black beard. His horse breathed easily as he swung down from the creaking saddle and led the horse to me with his right hand stretched out. "John Phelps", he said. We shook hands. "Nicholas Sadleir," I said. "I received your letter. I arrived this morning. Mr Chadwick has made me welcome, he was showing me the homestead when you arrived and he has returned to help the cook unload some wood and prepare for dinner."

John Phelps led his horse into the yard, replaced the slip rails, unsaddled the horse, removed the bridle, freed the horse with a pat and carried the bridle, saddle and saddlecloth through a small opening in the yard into the shed where he hung the saddle on a rack jutting from the wall and the bridle on a hook beside it. He spread the saddlecloth over the saddle. All this he did in silence. Then he turned to me, standing straight and taking me in as if appraising me. I guessed him to be about 40, tall and self-assured. "What news of your mother, Nicholas? And of your father James as well? I met them years ago at Willowbank, our home in Limerick before I left for New South Wales, resulting from a summer boating holiday your parents took on the Shannon." He retained the accent and manners of Limerick man. Perhaps it was to make me welcome.

"My parents are both well. I am one of three brothers who landed in Melbourne about five years ago. We regularly circulate letters from home. They report things much improved since the famine, although some of the big houses and the large estates are disbanded. I expect you, Mr Phelps, are well aware of the large numbers of Irish in Australia. Even more went to America."

"Please call me John, Nicholas. Phelps is too formal a tone for cousins, even if we are distant cousins. In fact, I am unsure of what kind of cousins we are – third cousins? Or cousins several times removed?"

John Lecky Phelps – : Photograph Courtesy Clare County Library

"I know not either, John. My mother never told me. But I believe we are related through one of her father's sisters who married a Phelps. There is a line of Clarke-Phelps in Galway I think."

"We are far away now. I expect the people and the affairs of the colonies of New South Wales and Victoria take most of our attention. As you may know Nicholas, I, like you, arrived in the colonies with two brothers. We have assisted each other in various ways. Joseph and I are both graziers. We began in the Wellington district, much closer to Sydney, about 20 years ago, learned something of the management of livestock and people in Australia and have been moving west to broader horizons ever since. Brother Robert has been involved in grazing as well, but he is the farmer of the family and has a vineyard at Albury up river on the Murray. He came there recently from South Africa."

I told John of my brothers Richard and John, Richard in Melbourne establishing an accouchement hospital, and John with the police at Beechworth, and we exchanged news of events from our counties in Ireland. I omitted news of our Parliamentary banker cousins and their disgrace in Tipperary, he may have known, but I did not seek to remind him, nor did I talk about my adventures on the goldfields. I was cautious of my Quaker cousin. I was unsure of what he would approve or disapprove. I had befriended ex-convicts, Roman Catholics from continental Europe, Hebrews from London and Chinamen from Canton on the goldfields, Blue, but I was still unsure of a countryman who was a member of the Society of Friends. I knew nothing of the Society, except that some members of my extended family in Ireland would adopt a pained expression when members of it were mentioned. Some, if they spoke of Quakers at all, called them *decadent*. And I knew little about this new cousin. But what I could see of his bearing and manner made him splendid. It was something about the way he handled his horse and his straightforwardness with me. He was obviously Irish but neither of these things reminded me of Limerick or Tipperary. He had evolved into something almost heroically special. I began to hope it would happen to me.

The cook arrived from the house in a dray pulled by a roan mare. He was Chinese and the carthorse mare had one eye missing. The cook seemed nervous with the horse so John and I helped him back it and the dray into the shed, unhitch the dray, unharness the horse and eventually release it into the yard.

But we had some fun before that. Ah Chee, the cook, left the horse to me (I who happened to be holding its fractious head) as soon as we released it from the dray. He ran to the storeroom and peered around the door jamb as I hung onto a rearing carthorse which tried to strike me with its front feet. I avoided the feet by lifting myself onto its back and taking the reins firmly. The mare reared and pig-rooted and I held on with gripping legs and a hand on the harness.

She bucked, backed and reared herself to an exhaustion we shared at the end of battle that lasted for at least 10 minutes. When she stopped misbehaving, I drove her forward with my heels and we cantered and trotted away from the yard and towards the river. I drove her down a shallow bank and into the river where she plunged again in fear and excitement. Eventually she stopped and drank. When she finished I guided her up the river bank and rode her quietly at a walk back into the horse yard. John fitted the slip rails and I unharnessed her as she stood quietly and permitted me to do it. I shook as I carried the harness to the shed.

Soon after, I vomited. I'd injured myself Blue, in a way that is impolite to describe accurately. Ah Chee emerged from the door and lowered his head with embarrassment. John looked at me levelly. "I can understand your pain. It's happened to me, but well done. You preserved yourself and a useful horse, and I liked your kindness in taking her to water after you had punished and subdued her. You're brave and you seem quite skilled, Nicholas. Would you consider giving her some more education? Don't answer now. I certainly don't mean today – you need time to recover." And he put his arm around my shoulders and he and I and Ah Chee walked towards the homestead. When I paused to stoop in pain holding onto a sapling, the other two paused, waiting respectfully. John said "Ah Chee, this is Nicholas, a cousin from Ireland, Nicholas, Ah Chee." Ah Chee bowed, averted his eyes and then peeped back at mine. I looked at him and grinned and then we giggled and the laughter became infectious. John joined us. Quakers could laugh too!

I've forgotten the pain but I can still remember the elation of feeling respected by a man I felt in awe of. And it had happened accidentally as a result of a troublesome horse. For the rest of his life John Phelps and his brothers taught me all they knew about pastoralism and commerce.

Nicholas Chadwick walked out from the house grinning. "I saw something of what happened from here," he said. "Ah Chee and I didn't have to unload the fire wood. The horse did it for us. She reared and the wood slid out of the back of the dray. Mind you the wood is away from the house and I don't think Ah Chee enjoyed the trip back to the horse yard much, but gathering wood is part of his duties as a cook." And he grinned at Ah Chee. I doubt that Ah Chee shared his humour, but he served lunch – cold boiled mutton with a delicious relish that tasted of cabbage, onion and ginger washed down with black tea and damper and johnnycakes. I had little appetite but that which I tasted was good.

After lunch John Phelps led me to a bed away from the dining area saying, "I think you need to rest, Nicholas. I had hoped to ask you to come with me to look at a flock of ewes and lambs but we can delay that visit until tomorrow. See if you can sleep and we can discuss your future in pastoralism tomorrow. Relax and think about the things you want to know."

Ah Chee came to me later with a potion to relieve my pain. I drank it. I slept until dawn the following day. I think that may have been my first opium, Blue.

I declared myself fit to ride the following morning. I wasn't. I still felt pain, but I thought it would not be much worse on horseback. That was a mistake, but I've made worse, Blue.

There were five for breakfast – Ah Chee served us and ate with us: Stanley, the horse tailer, John Phelps, Nicholas Chadwick and me. We had warm damper and boiled salted mutton with sweet black tea in silence.

Stanley left after gobbling his breakfast to go to the horse yards to bring hacks to the house for John Phelps and me. I ran after him asking if I could help. He looked at me shyly and agreed. He was little more than a boy and small enough to be a professional jockey. He loved horses. He told me that Nicholas Chadwick had met him a year ago on a trip to Balranald near Cannally station (John Phelps' home). Stanley was the third son of a large family that cut and sold firewood for paddle steamers on the banks of the Murray. He met Nicholas Chadwick at the wood heap his family had made as the steamer pulled in to take on wood for its boiler. Nicholas was a passenger and Stanley's father had asked him about work for Stanley with horses. "Stanley is keen on horses, Mr Chadwick. I'm afraid he'll not get much experience with them with me. Mine is all axe work and Stanley doesn't seem to have the build for it. He dreams of being a jockey, his mother is against it, she thinks he will fall into bad company, so perhaps he can find a position with a gentleman like you."

"Mr Chadwick offered me a job straight away. He gave me the horses to look after when he and I and a blackfellow took a flock of ewes and lambs from Cannally to here. When we got here he asked me to stay as the horse tailer. I've been here for more than a year. It's been a good job but I miss Mum and Dad and my brothers and sisters, but I might see them soon. Mr Phelps and Mr Chadwick tell me that there might be another horse drive soon and that I might be able to go on it."

We got to the horse yards. The one-eyed roan mare I'd fought with the day before stiffened and raised her head. I moved to approach her but Stanley intervened. "I heard you were good with her yesterday Mr Sadleir, and from what I hear you might get her right, but she is a bad bugger. You're still a bit sore. Don't take her on if you're crook[6]. Wait a few days. Here is a hack for you today," and he pointed at a heavy grey mare with a dished head like an Arab. "Catch her, and I'll get Mr Phelps' usual hack."

I did as he asked. The one-eyed roan mare watched me with her ears flattened along her neck, but she made no move towards me as I caught and bridled the grey mare and led her to the shed to saddle her. "Call me Nicholas," I said to Stanley. "I guess we are about the same age. I'm 23. How old are you?"

"Fifteen," Stanley said. "But I suppose I have some advantages over you. I was born in New South Wales. I know the country along the river. You come from Ireland don't you? "

"Yes," I said. "So do John Phelps and Nicholas Chadwick. "

"Really? I think of them as being here always."

"I hope you will teach me something about this country, Stanley, and thank you for the advice about the roan mare. I will do some more work with her, but later on," I said and I mounted the grey and led John Phelps' bay to meet him at the house as he finished breakfast. The saddle gave me scrotal pain, but it was bearable.

John Phelps and I talked for hours that day as we rode out from the river to meet a shepherd about five miles from the homestead.

---

6. Crook means sick or ill

John told me the shepherd was a ticket-of-leave man. He was a convict assigned to his brother Joseph on a run Joseph had at Wellington closer to Sydney town. John didn't tell me the man's name but he said the man had taken an aboriginal wife in the Wellington district, and when he gained his ticket of leave and was permitted to move to another police district, he had brought his wife with him first to Cannally and now to Tarcoola. "I think there was some trouble with the Wellington tribe and they were pleased to have somewhere else to go," John told me. "In this country mixed marriages are inevitable. Joe and I are pleased to have helped to make this one successful. This chap has been with us for 15 years or so. He seems to prefer the isolation that goes with the job and the six children help with the flocks.

We talked about the aboriginal people. John asked me about my experience of them. I told him I had very little. "I met a few only on the fringe of the towns when I was gold-mining. I can remember thinking they thought our search for gold slightly ridiculous. I can remember one old man sitting on the front of the store throwing a handful of nuggets on the dirt road at the front and roaring with laughter as diggers scrambled to pick them up. After a while he just turned his back and walked away.

A few I met could speak a bit of English, but they didn't seem to want to join mining or have any regular kind of employment. Sometimes they would beg for food or trinkets, but they were not like beggars in Ireland, they seem to have no shame in asking for things.

I expect my most concentrated contact with blacks was with Sammy just recently when he brought me here from McLeod's Crossing. He showed me an aptitude in tracking and hunting I had never seen before, and I met some members of his family who showed me some primitive methods of preparing food that tasted delicious."

John kept silent for a minute or two. "Perhaps I should give you a broader view, Nicholas. My brothers and I have been fortunate in that we have interacted with blacks in several shepherding districts. At first we thought all black people were the same but they are not. They look the same, but from district to district, they have different languages and customs. At first I hoped to introduce Christianity and schooling to the people I met with my flocks. The people met me courteously and they appeared to learn things quickly, but I think it was simply out of politeness. I became frustrated with my efforts, but when I reflected on it, I concluded that many of the things I offered the blacks in the various districts were not pertinent to their lives. We value wealth, fine houses, fine clothes, books and works of art and we work hard to strive to get them and keep them. They do not. This does not mean they are inept or dull. If anything, it is quite the reverse. They are curious and learn things quickly and they seemed more athletically skilled than we are. But I've yet to discover the higher-order things they really value.

In an ordinary way, they value the things we value. Parents love children. They nurture them. They teach them language. They teach them the skills of hunting and tracking from the time they can walk. And you may have noticed Nicholas, these are extraordinary skills. Five-year-old children recognise the footprints of their parents and their brothers and sisters in places with footprints of hundreds of other people."

"The people appear to be primitive John. Is there no underlying law, custom or religion governing behaviour in the way that the Queen, Parliament, the law and the church influence our lives?"

"The older people appear to govern Nicholas, and there does seem to be a rule of law because punishment for wrongdoing can be quite severe. My difficulty is that I do not know what their laws are (and I should return to my warning that we should not be generalising about groups in differing districts) but given that caution, I do not, for example, believe that blacks regard murder as seriously as we do. Murders of blacks by blacks appear to go unpunished. When we hang blacks for murder, their relatives are at first bemused at the ceremony we make of trial and execution and then outraged at the death of a clansman in the final result. It seems they may not have even made the connection between the crime and the punishment.

And crimes against property seem non-existent for them. I suppose it makes sense if you believe they own nothing, or that individually they own nothing and that everyone, communally, owns everything."

"May we translate that as meaning blacks own the land communally?"

"No. We may not. The Crown owns the land and we, as leaseholders in these marginal districts, pay rent for it. There is no question that blacks own land. They roam it. We the pastoralists are obliged to respect their customs and allow their hunting, and our police are there to protect them as well. The blacks have an obligation too. They must leave our stock alone. They must allow stock to water undisturbed. When tribes understand and respect that, there is harmony. Joe and I regularly employ blacks for seasonal work – sheep washing, shearing and yard building. They are good workers but they do not usually commit to a task for an extended period."

We talked of the rivers. John told me that since he had been in the district the rivers had never run dry (there were always pools and lakes) and that every five years or so there was a flood when the rivers broke their banks and water flowed for miles. "These rivers are the backbone of this country Nicholas. This is the spine of an overland stock route from Sydney to Adelaide and I think the Riverina (as some people are starting to call it) will be the permanent home of woolgrowers in New South Wales and that we will establish a food bowl of fruit for sale to the world irrigated with water from the river as well. Furthermore, the paddle steamer trade will make these rivers the highways of New South Wales."

"Even on this, the Darling?"

"Yes, Nicholas. Not every year, but when it is navigable. We have yet to know how far upriver steamers will go. The Governor of South Australia has already visited our place at Cannally on a steamer that came up from Goolwa in South Australia. All the colonies support the development."

'My head was spinning, Blue.'

'What about John's assertion that aboriginal people owned no land?'

'Oh, I accepted that. What I came to a few years later was much more puzzling; some of the Barkindji whom I worked and lived with told me that they belonged to the land. It was quite a different precept to anything I had ever learned.'

'This may interest you, Holas. In 1993 the Parliament of Australia passed the Native Title Act. It said terra nullius was wrong (the idea that no one owned the land before the Crown took it for allocation to its citizens, subject to colonial rules – that which you and John Phelps believed) and that aboriginal people had title to their land.

'What! That would have destroyed orderly governance.'

'No. There were rulings that if the title had been extinguished by the Crown then native title no longer existed.'

'What about pastoral leases? What happened to those?'

'You have not lost your acuity, Holas. There was a special High Court hearing about it and the court ruled that if an indigenous group could demonstrate a long-term attachment with the land, then native title could be granted. But, before you ask another question, the granting of native title did not cancel the grazing lease. Enactment of the act depended on the two claims coexisting.'

'That sounds rather like the philosophy John Phelps expounded on the day I went riding with him with my injured crutch.'

'And it seems to take into account the observation Barkindji people made to you that they belonged to the land.'

'Yes it does. What happened as a result of the change to the law?'

'Hardly anything, Holas. It's been mainly symbolic. Some aboriginal groups now own large tracts of land but that hasn't come about as a result of the Native Title Act. Most of the pastoral leases were purchased by government for Aborigines and money came from mining royalties too. What may interest you particularly is that the small homestead block of Weinteriga just across the river from Albemarle now belongs to an Aboriginal community in Wilcannia.'

'And that was all the land for the Barkindji?'

'Yes.'

5

Head Stockman

John Phelps makes Nicholas Head Stockman. That disappoints Nicholas. He hopes for a partnership, but blossoms in the role. He rides to Wentworth with a team to buy stores. He continues by paddle steamer to Cannally station and thence to Valverde near Albury, delights in life in a civilised household, selects horses and joins a team to walk them to Billilla on the Darling. The drive takes three months. Nicholas learns from an Aboriginal stockman and takes charge of the drive. He builds yards for a horse auction at Wentworth. He dismisses a roguish auctioneer and runs the auction himself. He is elated with its success and delivers more horses along the Darling River.

Great-Grandfather told me about his emerging understanding of the country and the way it was organised.

'It was early days at Tarcoola for the partnership of Chadwick and Phelps. They had only two shepherds, less than 3000 sheep and about 10 horses on Tarcoola, but the firm had grazing licences along the rivers from Cannally near Balranald to Billilla on the Darling south of Wilcannia. Cannally was the headquarters. Joseph and John Phelps and Nicholas Chadwick had a Merino stud there. As they improved the wool and size of their sheep, they sent breeding flocks with their improved rams out with drovers to their runs along the rivers. It was a central system of colonisation. They established each run with new shepherds with flocks, horses, a station headquarters, a store, tools and a system of communication and cordial relations with the local blacks.

Each day I rode out with John Phelps or Nicholas Chadwick to see to the flocks in the charge of the two shepherds. We rode wide of the flocks to find new grazing and we directed the shepherds to the new pasturage. Sometimes we helped to move the sheepfolds and the shepherds' camps to new pastures, but they were never more than a mile or two from the river – any further would have taxed the sheep and their shepherds; flocks went from their fold to water, to pastures and back to their fold at least once a day and sometimes twice. John Phelps told me that in some winters sheep did not need water at all. They drew the moisture they needed from the grass they ate and so shepherds could take their flocks further from the river. That presented another task in carting water to the shepherds for their normal sustenance and for their dogs. John told me that the people at the main station usually did that using casks containing water on drays or they used special moulded metal water tanks fitted to pack horse saddles to take the water out from the river to shepherds' camps.

We were able to use the one-eyed roan mare and dray to move the shepherds camp and sheepfolds by the end of my second week at Tarcoola. Each morning I had worked with her as a means of her civilisation. By the end of a week she was no longer dangerous. I rode her at first with a saddle and bridle and contrived an array of ropes connecting her legs that made her fall each time she tried to strike me when I was not safely on her back. Within a day the dangerous habit was gone and I was able to proceed with her normal education as a carthorse. She became a valuable asset – a tireless worker, but never a pet. She was always spirited and needed firm control. Ah Chee avoided her always. For all the time he was at Tarcoola as a cook, he carried firewood to his cooking hearth on a carrying frame he fitted to his back like a rucksack – no horse and dray for him! Stanley delighted in teasing him about it almost every breakfast time, "Ah Chee, you wan horse n dray for firewood? I bring up roan mare and dray d'rectly?" And Ah Chee always threw a pot at him. We seldom had dreary breakfasts.

John Phelps and Nicholas Chadwick continued my education. We looked at sheep, examined their wool in detail and compared the best with the worst. We talked about the organisation of shearing (it had recently been completed on Tarcoola, supervised by Nicholas Chadwick and done entirely by local blackfellows in bough sheds with temporary yards about 10 miles from the homestead). They talked about the provisioning of their stores on their runs, the cost of freight, the means of getting wool to markets, the relative merits of drovers and the importance of station cooks.

They were generous to me with their shared knowledge, but perhaps their instruction was to their mutual benefit. After I'd been there three weeks, after supper, Nicholas Chadwick and John Phelps asked me to walk with them to the horse yards. Nicholas lit his pipe (I had not yet started smoking, and I believe John Phelps never did) and said. "John and I have been talking about you, Nicholas. We like the cut of your jib. We understand that you want to invest in this country, but we do not believe that you have sufficient funds to run an independent affair, nor do we feel (and I'm sure you will agree) that you know not enough about the country or the management of livestock and the people of this colony to succeed in these marginal lands. Given that, we would like to employ you in a form of apprenticeship. We seek no money from you for formal indentures. We offer you employment, acknowledging your status and ours as gentlemen, at the rank of Head Stockman. For the time being you will not permanently live on any of our runs. Rather, we want you to share in the establishment of new flocks on new runs and we want you to take charge of the education of horses of all our runs along the river. For our part, we offer you a salary of £50. We will provide all accommodation, sustenance, tools, horses and saddlery and pay other expenses. We also undertake to instruct you in all aspects of the husbandry of sheep and cattle and in the proper conduct of sheep and cattle stations." He paused, settled more comfortably against the rails of the yard and looked to John Phelps.

John continued. "I feel obliged to say Nicholas; we are not being entirely charitable. I know we are distant cousins, but I believe we bear that as a faint obligation. In truth, Nicholas Chadwick and I believe you will assist us. You demonstrated ingenuity and courage by electing to walk here from McLeod's Crossing with a strange blackfellow. You have shown us your skill with horses. We are impressed with your cheerful balance of attention to detail and a broad appreciation of the relative importance of things. We like the way you have befriended people on the station and your fair dealing with them. The agreement we offer, we think, furthers profit and comfort to both parties."

I was speechless. If somebody had proposed marriage, Blue, I could not have been more flustered. I looked down, sat on my heels and doodled with a stick in the dirt. Eventually I found my tongue. "Thank you, gentlemen. Forgive my lack of composure. Please understand this is my first formal offer of employment in any country or colony. So far I have laboured in goldfields on my own account or with partners. I am unsure of my obligations to you as employers. Thus I do not believe I can fulfil them."

John responded after a pause for thought. "Your response persuades me even more of your suitability, Nicholas. You do not claim to know everything and by saying so you indicate your willingness to learn and adapt. Neither Chadwick nor I, nor my brothers, know with certainty what lies ahead of us, but we had even less knowledge when we began in this colony so many years ago. I will not beg you, but I respectfully urge you to join us."

When I reflect on it – at the back of my mind I had a nagging disappointment at their offer, Blue. Several days before, I remember telling Nicholas Chadwick I had £1,600 banked in Melbourne, and as I told him, I demonstrated I had money to invest without saying so directly, and so, as a result, I had hopes of the two offering me a share in their partnership. I still had the timidity of youth, Blue. I thought it was bad form to come straight out with, and it may well have been. To some extent we were still using the manners of old Ireland; brash youthful ambition was not encouraged, and I was their guest. However, they had dealt with the money I had in a straightforward way. A partnership was not what they offered but their offer of employment came with blush-making compliments. I hesitated no more. "Thank you," I said. "May I start work immediately? What plans do you have for me in the months ahead?"

John answered. "We have a large affair with horses planned, Nicholas. I want to take you and Stanley to Cannally to appraise the horses we have there, choose those that are suitable for our runs on this part of the river, draft them into a separate lot and leave them with Stanley to break and educate those that need it. When Stanley has his work set at Cannally, you and I will set off to my brother Robert's establishment at Albury. Robert has been buying horses for us. We will collect them, recruit droving staff, bring them to Cannally, sort and draft them again, work with those that need to be educated and set off with a drive of up to 300 for our stations on this part of the river and for sale at McLeod's Crossing."

"This is the drive of horses people in McLeod's Crossing expect in October, John?"

"Yes it is, Nicholas. But it will be late. We may have them there in time for Christmas."

"When we do leave? What must I do to prepare for travel?"

"Bring all you have. I hope we can leave at first light tomorrow. Nicholas Chadwick will remain here to attend to the shepherds, and you, Stanley, Ah Chee and I will ride to McLeod's Crossing. It should take us two days. Ah Chee will return here with the horses and saddlery and the stores we will purchase for him for this run and others along the river. Ah Chee is a fair horseman and he should manage alright – he just has an allergy to one-eyed roan mares. He will have three spare packhorses – the horses you, Stanley and I use. We can modify our saddles to become packsaddles, we will help him load the horses and he should be back with the stores from McLeod's Crossing inside four days. Nicholas Chadwick should have to put up with his own cooking for no more than a week."

"And how do we get to Cannally?"

"On a paddle steamer. There should be some coming up from South Australia looking for their second load of wool. I doubt we will wait more than a few days."

"And the same for Albury from Cannally?"

"Yes. With a fair degree of reliability. River traffic is increasing all the time."

The two older men looked at me. Nicholas Chadwick said. "Well what do you think of that, young Nicholas?"

"I'm not sure if I can think of a better way to spend the time up to Christmas. Which horses shall we take?"

"You choose. You're the head stockman."

That night John Phelps kindly provided me with pen and ink and I wrote home to Brookville House telling them my news. There was a post office at McLeod's Crossing to receive and send mail for England, Wales, Scotland and Ireland. I told my family of my new address and what I hoped to do with my life after goldmining and I sent this map to help them understand the spaciousness of my new surroundings.

And I wrote a short note to Richard in South Yarra, telling him briefly of my start in a new life and suggesting that he send letters to me care of the post office at McLeod's Crossing.

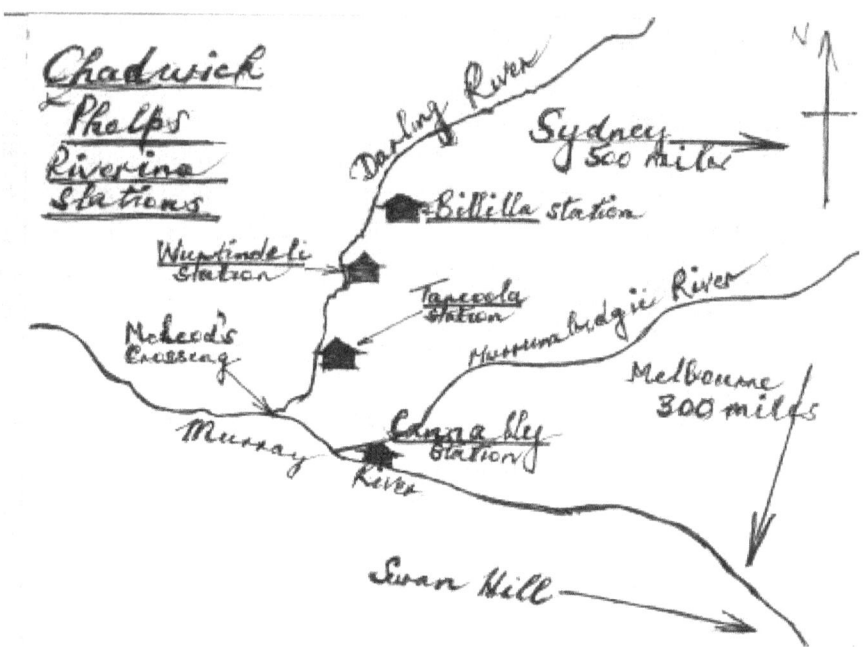

Map sent to Brookville House

John Phelps, Stanley, Ah Chee and I had a pleasant and rapid trip to McLeod's Crossing. We had a horse each and a packhorse to carry food and cooking equipment. We camped only once on the way beside the river. We carried our bedding individually and we carried hobbles for the horses on the packhorse. We hobbled their front legs to keep them from straying too far from our camp and we were able to catch them easily in the morning. It was an uneventful trip. We saw no blacks, but we did see signs of recently abandoned camps close to the river. John Phelps told me he thought they were fishing and duck hunting camps. He said the blacks often feasted on duck and fish at this time of the year. They had no way of preserving meat so they simply ate it excessively when it was there.

It was a hot late September day when we got to McLeod's Crossing in the mid-afternoon. Stanley and I tailed the horses with their bridles removed and their saddles still in place along the river about two miles from the township and we used hand axes to cut saplings into three-foot lengths to lash to the riding saddles to make them packsaddles for Ah Chee's journey back to Tarcoola. We had instructions to return to town with the poles within two hours. John Phelps and Ah Chee went immediately to a providor and bought flour, salt, preserved ginger, tea, pepper, dried apples, dried vegetables, potatoes, currants, raisins, sugar, mustard, vinegar, curry, rice, essence of lemon and pepper.

I estimated we had four hundredweight of stores to load on Ah Chee's caravan of horses. We lashed the makeshift poles we had cut from saplings to the saddles with rawhide thongs, attached empty flour bags to the poles on either side of the saddles and filled the bags with stores. It took the four of us more than half an hour.

"I don't envy you loading and unloading the horses on the way home to Tarcoola, Ah Chee," I said. "But if you catch and load each horse one at a time, it should be pleasant and tranquil work. And whatever you do, don't forget to hobble the horses at night."

Stanley interjected. "My word, Ah Chee, if you lose the horses you'll have to walk to Tarcoola and come back with the roan mare in the dray to collect the stores. Have a care!" And Ah Chee paused; turned away from the horse he was packing and shook a stick at Stanley. But he was grinning.

Ah Chee rode off leading his three packhorses in a string about an hour before dusk. He told us that because the horses had been well fed before he left, he would tether them at the first camp and not release them to pasture in hobbles to save time on the first day by not having to walk out to find each horse. He expected to use hobbles after that. He was a good horseman. I asked John Phelps, "How did Ah Chee find his skill with horses?"

"Brother Joe first employed him at Wellington probably about 15 years ago. We think he came up from Victoria but his history is a bit hazy. We took him on as a cook and gardener and when we moved from Wellington to Cannally he came with us as we walked our cattle, horses and sheep west. He learnt about horses on the road.

He's become a very good all-round man. I hope he'll be alright on the trip."

"What do you mean?" And I turned to him.

"There was news of a hut-keeper further east who was speared. There is always talk like this in the towns. I had a quick word with the police. They tell me it is an isolated instance. They almost implied the hut keeper got what was coming to him, he had been entrapping black women, but it is always a little worrying when spears come out against us. Ah Chee should be alright. He can be quite the diplomat. Usually he cooks his way through trouble, and he is armed. He has an old muzzle-loading carbine with him if the worst comes to the worst."

Nicholas talked on:

'There were no steamers heading up river. We went to a newly opened hotel for supper and immediately lamented Ah Chee's departure – tough salt beef and damper only. John Phelps demurred at the expense of a lodging house, so we slept in the open on the banks of the river.

The Bluebird tied up to the wharf with a barge in tow. I watched it approach and was delighted to see Sam Curtis with his shock of curly grey hair emerge from behind the boiler. "Nicholas Sadleir, even if you are the brother of a trap, it is a joy to see you."

I stared at Sam. "I didn't expect to see you for years after our time at Swan Hill, Sam. Where have you come from? What have you on board? Where are you going next?"

"I've just returned from delivering a load of wool to Goolwa, Nicholas. We called at a few stations downriver with stores on the way here, and we have a big load to deliver. The barge's fully laden with goods for several at McCloud's Crossing. We have about five ton of flour, wine, dried fruit and lots of haberdashery and manchester. Enough to start an emporium. Our next delivery is at Cannally – general station stores for Chadwick and Phelps, they are regular clients of a providor in Goolwa. He sends up an order about every six months.

Anyhow, you ask more questions than your trap brother does. It's time you answered a few of mine. Obviously you got to McLeod's Crossing from Swan Hill because here you are. I'm anxious to investigate your mischief and find an explanation for it. Have you been up the Darling?"

"Sam, I believe you're trying to be a policeman. Remain as a boatman. I have progressed. I think it's highly likely that I may superintend the unloading of your goods at Cannally."

"What in the hell are you talking about Sadleir? For you to "superintend" (and what the hell does that mean?) you will need to be at Cannally when I arrive. And that will not be long. You need to get a move on."

"But I may come with you."

"Why do you think I will permit you aboard?"

"For friendship. But I may need to protect you commercially, Samuel. Lower your voice. That tall man watching us is a partner in Cannally. He now employs me. You have goods belonging to him for delivery to one of his stations. He is establishing many more along these rivers. He will need wool floated to ports to join with ships bound for England. He wants paddle steamers to do it for him, but if you behave foully, he will choose others. Many more steamers are needed. You have some acuity, Sam. Use it."

Sam stared, blinked and nodded. He said (loudly enough for people on the shore to hear). "I'm delighted to see you again, Mr Sadleir. I understand you are bound for Cannally station. How many are you? May I be of assistance?"

And so we secured passage to Cannally. Sam Curtis and I leapt from the deck to the bank and walked to John Phelps. John eyed him warily but Sam was on his best behaviour. After I'd introduced him to John, Sam said "I'm very pleased to make your acquaintance, Mr Phelps. I met Mr Chadwick here about a year ago. I like to know the people I work for. Nicholas tells me you seek a passage for him, yourself and a horse tailer to Cannally. I am pleased to offer you a cabin and sleeping space for the other two in the mess room. I assume you have your own bedding? I will make no charge for the passage. I'm honoured to carry your goods and the transport of your good self and members of your firm will embellish the honour."

Sam startled me with his oily eloquence but John just nodded with a wry smile and accepted Sam's offer. I think he may have been making a mental note to compare Sam's prices with those of other riverboat men. I feared Sam might have overdone things. I think he may have still been thinking like a convict. I told him as much when we drank rum leaning on the bar in a tavern facing the junction of the two rivers that evening after John and Stanley had bedded down on the Bluebird.

"Sam you seem too eloquent and too servile in the presence of John Phelps."

"Wadya mean?"

"You don't appear to be treating him as an equal."

"Well he isn't. He's my master. And I thought I was talking to him in the way you wanted me to."

"I need to tell you more about him, Sam. He is a plain devout man. He comes from a sect in Limerick who call themselves the Society of Friends. There is nothing ascendant about them. They are plain hard-working God-fearing Christians who seemed to turn a profit at most things. Servile flattery will get you nowhere. I think when you were buttering him up with an offer of free passage he was surmising that perhaps you are charging too much for freight from Goolwa to Cannally because you offered to carry us for nothing. He is a good man. A plain man. You are my friend. I trust you because I got to know you well at Swan Hill. Be with him like you were with me. Be yourself. Respect him but allow him to respect you. When you talked like that I knew you were pretending but I think he did too. That long nose of his isn't just for breathing."

He stared at bottles standing on shelves behind the bar for about a minute without speaking. I feared I had offended him. Then he turned to me and spoke. "Thanks. It is easier for me to be me. What should I do?"

"Talk to him frankly about your plans in the riverboat trade. Ask his advice. Hold nothing back about yourself. He worked with convicts in the Wellington district. He treated them well. Tell him your entire story, about your sister and about the farm at Echuca. But most importantly Sam, listen to him. He has some inspiring visions for this region he calls the Riverina."

I felt embarrassed giving the sermon, I was Sam's junior by some years, Blue, so it ended there, but I think it worked. On the voyage to Cannally Sam and John talked. Sam could read and write which was unusual for a convict, and he and John pored over maps and charts of the river systems and wrote calculations about the costs of operating steamers. Sam operated a string of steamers for years along the Murray, the Darling, and the Murrumbidgee. I think his talks with John on the way to Cannally helped him decide how to do it. He was also one of the first to go up the Darling from McLeod's Crossing. I'm sure John Phelps had a hand in that.

The banks of the river slid by. It was as beautiful as I remembered on my trip downstream from Swan Hill. I relaxed but Stanley couldn't. He stood at the bow straining his eyes up river looking for signs of his father's wood stacks. He had already spoken to Sam and the three other members of the crew about it. "Do you know William Williams' firewood stacks? Best wood, best prices. Probably about 30 river miles[1] this side of Cannally. It's my father's place. Surely you know it?" I saw the crew smirk at each other and look to Sam. Sam said, "Can't say I do Stanley, and anyway we should have enough firewood to get us to Cannally. I know the bloke cutting wood there. He usually gives me a good portion at a considerable discount."

Stanley was crestfallen. Sam's perverse sense of humour prevailed on the Bluebird. It wasn't until Stanley had sighted his family's firewood emporium and had started to move back towards the wheelhouse to plead leave to see his family that the steamer and its following barge slowed and moved towards the bank beside the wood heap without a word to or from the helmsmen. Straight-faced, Sam turned to Stanley. "I've heard a bit about this woodcutter bloke, Stanley. People say he is a bit rough. Nicholas tells me you are from around here. Have you any advice for me about how to handle him? Perhaps you should go to investigate to see if he is a fit and proper person for me to engage in a commercial conversation. We will just wait here for a while. Perhaps we will have a drink of tea." And he blew the steamer's whistle.

Stanley finally saw the joke. John Phelps smiled at him. "Go Stanley. You owe your family a visit. It is late afternoon. I think I

---

1. It was usually three times further to get to a place along the river than to go straight to it on land. All the rivers curled endlessly. Thus, river miles denoted extra distance.

may be able to persuade Mr Curtis to moor here for the evening. And I look forward to meeting your family." By then Stanley's family had answered the call of the ship's whistle. Mother, father, a mature pretty girl whom I noticed immediately, two young girls and two boys aged about eight and 11 came running down to the bank of the river. The crew were busy with ropes tying the barge and the steamer to trees beside the riverside stack of dried, cut and split red gum limbs. Sam slid out a rough gangplank to the riverbank. Stanley rushed down it carrying a sugar bag and was immediately mobbed by his family who were surprised and overjoyed. Stanley had bought presents for everyone at McLeod's Crossing. He embraced them each in turn and gave them their gifts. The two boys got pocketknives; his father a sharpening stone, his mother a box of needles and pins, the two girls brightly painted papier-mache dolls and his senior sister got a tasselled silk shawl. She swirled in it. Our eyes met. I hoped to meet her.

And eventually I did. After the present giving and a short explanation by Stanley about why he was here on a paddle steamer on his way up river instead of tending to a troop of horses, he led his family across the gangplank and on to the Bluebird. "Mr Phelps, I would like you to meet my family." And he introduced them each in turn to John Phelps, then to Sam Curtis then to me and then to each of the crew.

The girl was lovely. She was Rosie. In the long business of introductions, we seemed to be standing beside each other frequently, but it was difficult to think of anything to say.

William Williams, the father, made a speech. "Gentlemen we are honoured to meet you and thank you for bringing home our strong and healthy son. We thank you especially, Mr Phelps, for employing Stanley to look after your horses and in encouraging him to learn the trade. Will you please be our guests for supper this evening?" We looked to John Phelps. He nodded. "I fear we may be imposing on your kindness, Mr Williams. But we graciously accept."

Mrs Williams bustled. She addressed her family. "Come everyone, we have important guests to prepare for – some of them are sort of members of Stanley's new family. And she turned to smile to us standing on the boat. I will send one of the boys to fetch you when supper is ready."

Sam moved to William Williams. "We need wood for the boiler, Mr Williams, but I will not impose on your hospitality to gain it. I will pay you a fair price. This is your living. I have room for half a ton. May we load it now before supper to be ready for a start to Cannally first thing tomorrow morning?" Stanley's father agreed with a nod. "Have you any trading goods, Mr Curtis? Perhaps we may set the wood against some flour, sugar, tea and dried fruit?" "That will be entirely satisfactory, Mr Williams. Perhaps Mrs Williams has a list and one of your family may bring it to me so that I can prepare to carry it to your home when we go there for supper." The crew and Stanley's father started to carry the wood to the stack on the boat by the boiler.

Rosie Williams came with the order of grocery items from their mother. I hoped she would and she did, and at the same time, Stan Curtis asked me to weigh and parcel the items she sought from the steamer's trading store. We were awkward. Rosie could not read. She had no written list. She recited it to me and I wrote it down with a pencil on a scrap of paper, but when I turned the list to her to verify it, she blushed. "Please will you read it to me again, Mr Sadleir?" It was my turn to be flustered. I had embarrassed her and she was too beautiful to be embarrassed. And so I read her the list, item by item, and I looked to her for her response. Her eyes were greyish brown and she lost her shyness as we went through the list.

"Ten pounds of sugar"

"Yes, Mr Sadleir."

"Please call me Nicholas. May I call you Rosie? A bag of flour?"

"Yes, Nicholas."

"Four pounds of tea?"

"I think it was three, Nicholas."

"Three it is then, Rosie. A box of candles?"

"Yes, Nicholas."

And so we caressed each other with our eyes and held our looks for longer as we dawdled through the list of 15 items. Sam saved us from further affection by bumbling in with some empty flour bags and a cheeky look on his face. "Here are some sacks to bundle up your purchases, Miss Williams. Perhaps when you have helped pack them you may carry them home for her Nicholas?" He checked the list. "The value of these items does not equal the wood your father has sold me, Miss Williams. Nicholas, will you please add another bag of flour to the consignment?"

    I even felt romantic staggering up the riverbank with a bag of flour on each shoulder and watching the pretty sway of Rosie's grey skirt ahead of me as she carried two sugar bags of smaller items. The family's household was about half a mile back from the river and set in a thicket of trees I came to know as Black Oak or Belah. It was an open, large, living, cooking and eating area similar to the one I had just left at Tarcoola, but there were small log cabins made in the American style from native pine (laid parallel to the ground and joined at the corners with notches locking the logs together). As Mrs Williams scurried about her cooking pots, Rosie showed me the household. Log cabins were sleeping quarters, one for the boys, one for the girls (Rosie and her two small sisters) and another for Mr and Mrs Williams. A fourth cabin served as a storeroom. All the rooms had dirt floors but they were shiny, not dusty. Rosie proudly told me she and her mother made them that way from wet pounded ants nests. The whole camp was neat and tidy. It was prettier than Tarcoola. It benefited from a feminine touch. I was benefiting too until Rosie's mother intervened, Blue. "Rosie, I need some help. Please will you start laying the table for our visitors? Thank you for helping to deliver our dry goods from the boat, Mr Sadleir. We look forward to seeing you with the others in a little while. One of the boys will run to the boat to fetch you."

    And so I was dismissed from promises of further intimacy, Blue. I suppose it was a good thing. There were stations up river that needed horses and I had to get them there. Rosie and I hardly exchanged a glance when we had supper at her household later that evening. I guessed her mother had done more to her than send me away but she provided our ship's company with an excellent supper of boiled fowl, potatoes, cabbages and onion and a rich raisin and plum pudding.

    Stanley spent the night with his family and joined us at first light as the crew stoked the boiler to get up steam. By the time we got to Cannally Rosie was a fond, lingering, but distant memory.

Cannally was rather like Tarcoola, but there were several buildings like the Tarcoola homestead and a thriving garden watered from the river. There were no white women there although several black women helped with the cooking. Neither John Phelps nor Nicholas Chadwick had married, so Cannally was a masculine affair. Eight white males and a Chinese gardener lived there with a scattering of black families who lived in humpies by the river. And there were shepherds in huts out from the homestead.

John had been in discussion with Stanley on the morning's journey from Stanley's family camp. John described the horses Stanley was to collect and educate and would introduce him to the overseer at Cannally to make this possible. John had decided that he and I would go directly on the Bluebird with Sam to Albury. He was looking forward to seeing his brother Robert and collecting the horses Robert had purchased before the weather got too hot. We were at Cannally only long enough to transfer the stores from Goolwa to the Cannally store.

In less than three hours we left for Albury. Sam and the crew drove the steamer night and day, there was a full moon, we could navigate easily at night, we stopped for firewood only twice and we got to Albury inside six days.

Sam had stores to collect to take up river to the station where he would load wool for Goolwa, so John and I left him at the wharf and we hired a buggy driver to take us to Robert Lecky Phelps' place called Valverde. It was a new gracious homestead built of adobe and surrounded by beginnings of luscious gardens with young vineyards and orchards stretching to horizons in most directions. There were geese and ducks in the garden. Robert's wife lived there with their children.

It was like rejoining civilisation. It reminded me somewhat of the homestead on an Irish estate, but it had another feel about it as well. I learned later it was African. Robert had been a successful farmer and horse breeder in the colony of South Africa and had removed to New South Wales only recently. Mrs Phelps was colonially born there. I saw neat, orderly outbuildings with roofs thatched with wheat straw. There was a forge, stables, a cowshed, chicken coops, a piggery, a large vegetable garden, several large storerooms, sheds for buggies drays, ploughs and harrows, and haystacks.

Robert Lecky Phelps was a younger version of his brother John but he talked more. "I'm very pleased to meet you, Nicholas. My mother often talked of your mother in letters to us at the Cape. I think they wrote to each other regularly. How are your parents? We did have news of the terrible famine at home and I often greet countrymen in Albury – some free, some not so free, but nearly all are content to be here."

He took us to newly built visitors' quarters near the homestead. These were separate to the single men's quarters (they were about half a mile from the homestead and were self-contained with their own kitchens, cellars and butcher's establishment.)

"I will arrange for our manservant to bring you hot water and towels for a bath. I expect you are tired from your journeys. Rest. My man will call you for dinner in the homestead at seven."

A bath! What luxury. Since the goldfields, when I indulged myself with a weekly visit to public bathhouses, I had not been in the habit of bathing regularly. At Tarcoola, when I washed my clothes in the river I washed myself, and sometimes I swam – Sammy had taught me on the way – but it was an infrequent and irregular routine. John Phelps had the first bath and Collins, the manservant, cleaned the bath and changed the water for me. The warm water and soap felt like silk. I washed and combed my hair and beard. And I had the additional luxury of donning a fine suit of clothes from a Melbourne tailor I carried with me in my travelling trunk (I had retrieved the trunk from the police at McLeod's Crossing before we left to go up river). I almost felt like an Irish gentleman again when Collins called us to the dining room.

Josephine Phelps stood with her husband Robert and indicated our places at the table. We were a party of four with two children and it was a pleasant reflection on civilised life enduring in places that had passed me by. Until then, most of my dining in Australia had been in bachelor establishments – public eating-houses, men's clubs and most often in rough camps on the goldfields – or most recently at Tarcoola. This was gentle and heavenly. We had fine wines, courses of fish, pork, fresh fruit, even cheese, and the beautiful Mrs Phelps guided our conversation to matters more elevating than the supply of horses to distant sheep stations on the Darling River. We compared colonial progress in South Africa with settlement in New South Wales and Victoria; we talked about the abolition of slavery and its consequences in India and North America. We discussed the end of transportation of convicts to New South Wales, and we talked of the writings of Charles Dickens. *Little Dorrit,* Dickens latest novel, had just arrived in the colonies and Mrs Phelps passed her copy to me. "I think you will find some of the passages interesting, Nicholas." And she raised her eyebrows.

She puzzled me. When dinner finished I read by candlelight in my bedroom. I finished the book just before dawn. It was an interesting story of debt, murder, life in a debtor's prison, a fraudulent banker, true love and salvation and much more besides. Mrs Phelps was delighted and surprised when I returned the book to her at breakfast. "My, you do read quickly, Nicholas Sadleir. What did you make of it?"

"I enjoyed it Mrs Phelps, but I think when you passed it to me you suggested that there may be something there of special interest to me. If there is, I have failed to grasp it. Forgive me. I have not spent my last years in reading and scholarship. Perhaps it has dulled my sensibilities. Please assist me to the point you make."

"I believe you had relatives in banking in Tipperary, Nicholas."

"Yes." I responded. And then her idea emerged. Merdle, the swindling banker in *Little Dorrit* had many characteristics shared by the fraudulent Sadleir bankers who cheated many Tipperary residents of their savings. After some thought, I looked at her and smiled. "Now I see what you mean, Mrs Phelps." Tipperary has been the inspiration for another literary endeavour. "Is that what you meant? Merdle – the banker in the story – and his likeness to my cousins?"

"Precisely."

"How clever, but it may be a coincidence."

"Indeed, but it is instructive to think about it."

Conversation with Josephine Phelps enlivened me to the joy of fine books and feminine company, Blue, but I was not to enjoy it regularly for many years to come. I basked in it eventually with your great-grandmother, but at the time of which we speak, she was but four years old, living without a mother, in the care of servants in her father's household at Geelong in the colony of Victoria.'

'I have some more questions about that basking Holas, but go on with your story.'

'John Phelps, with his dour practicality reminded me of my main purpose at Valverde. It was clearly not literary. My dreaming indulgence in feminine company lasted less than half a day. Soon after breakfast, we, John Phelps, Robert Phelps and I, rode out to a horse tailer's camp further upriver to look at the horses Robert had bought for the river stations.

Augustus O'Malley tailed the horses with three aboriginal assistants. He was a squat clean-shaven man (except that he hadn't shaved for days and his face was covered with dirty grey speckled stubble). He looked about 50, and he was sitting on a log in front of the campfire in grey canvas jodhpurs, bare-chested, and bareheaded with braces slipped from his shoulders, peeling potatoes into a large black saucepan. He knew John Phelps. John told me Augustus had served in cavalry regiments in India. He came to New South Wales with the British military but he had been cashiered in Sydney for insubordination and gradually moved westward, establishing life for himself as a drover. John told me he was difficult to manage. "He is amenable to good work if he can decide how he does it independently. He will brook no supervision entirely, but straightforward contracts seem to work for him, for example, "This flock of 3000 ewes to be delivered in good condition on such and such a run by the end of such and such a month, at such and such per head. Nothing else. I suggest nothing. If he wants something he asks for it."

Augustus remained seated. He leaned back to look at us as we dismounted. "Good morning, John, good morning, Robert. And who's this young toff?"

I didn't feel I was a toff, but perhaps I was. I was well dressed in riding breeches, a waistcoat and knee boots – not appreciably differently dressed to the others – but my clothes were new. I approached him with my hand out. "Nicholas Sadleir," I said. He ignored my hand and remained seated. He looked down and continued to peel his potatoes. "Nicholas Sadleir, I must explain one of the niceties of the Australian bush. Never interrupt a man preparing dinner for his workers."

John and Robert held the horses and watched, smiling faintly.

I stepped back, folded my arms and waited. He finished peeling the potatoes, added water and salt, and put the saucepan to the fire to boil. Then he darted to a tent and emerged buttoning and tucking in a deep blue linen shirt. He lifted his braces to his shoulders.

"And now, young gentleman. Why are you here?"

"I'm a guest of Mr Robert Phelps."

"But why are you his guest? Why are you standing in my camp?"

"To inspect the horses Robert bought for his brothers for sheep and cattle runs further down river."

"And who has asked you to do that."

"John Phelps has."

"Why?"

"Perhaps you should ask him."

"You're a cheeky young bugger. Can you fight?"

"A fight is not what I've come for Mr O'Malley. If that is what you want, I will oblige you, but I doubt it will be a fair match. I am taller, heavier, and younger. Your injuries will not promote your continuing success as a horse tailer's cook."

"You insult me young man. I am in charge of this establishment."

"Then please show me that you are and how it functions."

Augustus O'Malley turned to John Phelps. "John, when will you take delivery of these horses and what is this young man's part in it?

"I will take delivery of the horses once we have inspected them and agreed on the ones we will take and the ones we will leave. Nicholas will assist in classing. We will take delivery once we agree a price with my brother Robert. We need your help with the classing and we need your help in recruiting two suitable droving staff to accompany Nicholas and me to a station near Wilcannia on the Darling. They can expect to be away for three months."

"And what is the good of this young pup apart from his insolence?"

"Muster the horses, Augustus. I think you will see."

"Only for you, John."

"No, Augustus. For Robert and Nicholas too."

Augustus O'Malley, John Phelps and I worked together uneasily in the horse yards near Augustus's camp – or at least Augustus and I did. John seemed unperturbed. Augustus refused to look at me as we worked as a sort of committee agreeing to *keep* or *cull* horses as we drafted them through a gate – but it wasn't a sober office-based committee sitting at desks, Blue. We three stood in the yard with the horses near the gate we used to keep or release horses we wanted while the blackfellows pushed groups of horses towards us for our evaluation. John Phelps had proposed a system (and there was no argument about it from Augustus or me) – the horses we wanted to keep were released from the yard to the care of a horse tailer who would keep them in a separate troop; cull horses were retained in another yard and horses that needed further examination for a decision remained in the herd we stood with in the main yard.

Augustus or I "proposed" or "objected" and John Phelps "accepted" or asked for further debate. John was even-handed. Conversations went something like this:

Augustus. "Keep her, a good young heavy mare. She should be a good breeder as well."

Myself. "Agree."

John. "Let her go to the herd we want."

Discussion on another: Myself. "Cull him. He has little work left in him and he has a damaged near hind fetlock."

Augustus. "Keep him. A good old horse with years of work in him yet."

John. "Leave him in the yard for another look later on."

And so it went on. By the mid-afternoon we had about 40 horses in the yard Augustus wanted us to buy, and I had declined. I knew I had to be on my mettle. We took the broken-in horses first and had a blackfellow catch them and lead them to us. John adjudicated.

"Why refuse this horse Nicholas?"

Myself. "I judge it to be 18 years old. It is past a useful working life. It may have two or three more years of light work. It is a gelding, so it has no use for breeding.

Augustus. "Rubbish. It is less than 10 years old."

Myself. "We need to examine its mouth." John and Augustus held the horse's head while I looked at its teeth. "This groove on this tooth suggests to me that this horse is at least 25 years old."

Augustus gave me a long stare. He chose another. "What about this one? How old is she?"

"I am unsure. We will need to mouth her."

"You can't. She hasn't been handled."

"Mr O'Malley, I think the reason I suggested we not buy her is because of her twisted off front fetlock. However if she is young enough, perhaps you have persuaded me to think again. I think the twisted fetlock came from an injury. It is not an inherited fault, so she could be good for breeding. She should breed good hacks. Without examining her mouth, I judge her to be between two and six years old. If she is six, she is not a good proposition, because not having a foal at foot means that she is probably infertile. If she is younger, she may be worth buying. What do you think? Is that why you asked me how old I thought she was?"

"Could be." And he grinned.

"If we can get one of the men to get a rope on her and tie her to quieten her for a while I think I can get a twitch on her sufficiently to keep her still and we can look at her mouth together."

Augustus nodded. We restrained the mare and examined her mouth in the morning. We agreed the mare was about three years old and jointly recommended to John Phelps that she be released to the herd we planned to take with us.

Augustus said to me. "Perhaps you can learn something, you young toff. I reckon I can show you a better way to use a twitch." And he showed me. It wasn't any better than my method, but it wasn't any worse either. We agreed on that.

Augustus and I worked well together in the dust, noise and sweat of horses after that, and I could see that John Phelps was pleased. After three days in the yards, he had a herd of more than 200 horses being tailed beyond the yard and he had agreed on a price Chadwick and Phelps would pay his brother Robert Phelps, and further, Augustus had agreed to contract for the drive to Cannally. He would take one blackfellow with him and had reduced his fee per head because John had persuaded him that he and I would work under his direction, at no cost to him. John Phelps certainly taught me the art of diplomacy and persuasion. He had a firm and gentle way with him.

We finished the work by branding the horses – the brand of J. L. Phelps for those we were taking with us and Robert Phelps' brand for those he would retain. He planned to sell most of them.

We had been two weeks in the horse yards and John Phelps and I moved to Augustus' camp with our bedding and luggage, ready to depart as Augustus decreed. As contract drover, Augustus provided saddlery, horse hobbles and tethers, cooking equipment, food and a cook, but in a departure from normal practice, he and John Phelps had agreed to use suitable horses in the herd as riding and pack animals and that we would educate *green* horses as we progressed. Augustus was able to leave his normal droving plant of horses at Albury in a paddock at Valverde with good feed and water. He expected to be away a month.

We left early in the morning in mid-November with 251 mixed horses – heavy horses for drays, carthorses and light hacks.

Augustus was in his element as boss drover and he quickly established his authority and a routine for us. We kept to the river for the whole journey. At first light, after breakfast, everyone would assist in catching the hacks and packhorses for the day – harnessing them and tethering them; and two would dismantle the temporary rope yards we used, pack the rope on two packhorses and set off down river to stretch another. The remaining two tended the herd as the horses went to water and then moved them to grass away from the river. The cook packed his kitchen, food and our bedrolls to two more packhorses to establish a halfway lunch camp. The two who went out to establish yards for the next night took a lunch of cold meat and damper with them. Those driving the main herd of horses met the cook in his camp, picked up their lunch and ate it in the saddle. Then the cook moved to find the new horse yard and set up camp again.

Without doubt, the cook was the most important member of the drive: Tom Watkins worked for Robert Phelps at Valverde as farm overseer, but he had never been down river and persuaded Augustus to take him on as a cook with Robert Phelps's permission. Tom treated it as a holiday and enjoyed it. We carried salt meat, and that lasted for the first weeks of the journey, but after that Tom hunted and fished for our meat. We had Murray Cod and kangaroo to go with our damper on most days and Tom made delicious johnnycakes with currants and raisins brought with us from Valverde.

Jacko, a blackfellow from the Albury district, worked with Augustus regularly and he and I formed a team of two. We rotated with Augustus and John Phelps – moving the yards on one day, taking charge of the horse herd on the following day, and then rotating again. We carried hundreds of yards of light rope. Jacko was excellent at finding groups of trees we could use to make temporary fences to form an overnight yard for the herd.'

'But surely a yard made of light cord and strung to trees would not have been secure for more than 200 horses, Holas.'

'I had my doubts as well at the beginning, Blue. But the temporary yards worked remarkably well as a bluff – we ran three taut lines, the highest was about four feet. The horses were quiet and we had someone riding around the outside of the yard all night. We all took turns at that – three hours at a time. The herd did break out once during a thunderstorm, but everyone was up and on horseback when that happened and we were able to keep the herd together. We kept three horses saddled all night and tethered near the camp for emergencies.

Thinking about it now, people wonder how we put up with it, it seemed such a lot to do, but in many ways, it was a relaxing routine. On the days we were stringing rope for yards, Jacko and I would often finish them soon after lunchtime and he would teach me to track, to fish and to hunt kangaroo. We often helped Tom out with fresh meat for the pot and have it waiting for him when he arrived to set up the night camp. Augustus and John Phelps were not quite so keen on fishing and hunting. They never said so but I suspect they used to have a nap in the afternoons. They were a bit older than Jacko and I.'

'What about the droving, Holas, how did you adapt to that?'

'Well it was certainly different from leading tethered horses along Tipperary roads. Mostly the horses made their own way, they just fed along, sometimes we would have to slow them down, but usually one of us would lead the herd and the other would trail along behind making sure the stragglers kept up. You got to know where horses in the herd would be and if you noticed them missing you would go to look for them. Overall, it was relaxing. Often Jacko and I would just let the herd look after itself and we would talk. He told me the name of plants and tried to teach me how to track, and I gave him some more words of English.

As well, we used to catch and educate young horses as we went along. We didn't ride them; we just got them used to having a halter and walking beside us.

Yarding the horses at the end of the day was never any trouble. We would take them to water along the river first. There were several horses that were always in the lead in the herd and we would throw a rope on a couple and lead them into the yard. The rest just followed. In the end we didn't even need to use a rope. The leaders would just follow us in.

By the time we got to Cannally I had learned to drive horses. Jacko and Augustus were worthy tutors. "For a young toff, used to the hunting pinks of Ireland, you haven't scrubbed up too badly, young Sadleir," Augustus told me as he bade me farewell just before he, Tom and Jacko boarded a paddle steamer to take them back to Albury with their saddles and bedding.

"Thank you, Mr O'Malley, and I thank you for engaging and instructing me in the rude systems of equine management of the colonies." We maintained our tense modes of address all our lives, Blue. He took charge of many flocks of sheep and mobs of cattle for me regularly for years. We were close friends, but we never said so.

Stanley was pleased to see us. He had 15 light mares, 10 foals and a thoroughbred stallion to join the drive. They were quiet and well mannered, newly branded with the Phelps horse and cattle brand and in good condition. We talked about arrangements for the next stage in taking horses along the river to McLeod's Crossing. John Phelps told me he planned to remain at Cannally. We spent two days in the horse yards making lists of horses and agreeing on the prices I should ask for them when we presented them for sale at McLeod's Crossing. We reserved the best, a balance of heavy and light horses, for the runs along the Darling. I recall we marked about 150 of these to go north with the breeding herd Stanley had collected for Tarcoola, we left about 50 good sorts for Cannally and kept about 70 for sale at McLeod's Crossing.

Stanley told me he had recruited a young blackfellow to help with the drive and sought my approval. I agreed. John Phelps had assigned three of the Cannally staff to the drive as well. I met them as they were drawing lots to decide who would be cook. Harry, the youngest, had drawn the short straw and protested vociferously. "You bastards, I can't even boil water, you will rue the day you did this to me." He pleaded. "Charlie, Hughie you can't do this to yourselves. I'll poison you." Charley and Hughie turned their backs on him and walked away. He turned to me and opened his arms as if appealing. I said, "Do you need me to teach you to cook, Harry?" He shrugged. "All right, here's a list of tucker we will need for the drove. Mr Phelps and I have calculated what we will need for two months. Take it to the station store and start assembling what is on the list. Put the tucker with your cooking gear in the shed beside the horse yards and we will help you load the pack horses with your larder, cooking pots and our swags." He shrugged again and wandered off towards the store with my list in his hand.

Harry turned out to be a good cook. On the first night out he gave us yeast bread he had persuaded the Cannally cook to give him and we had roast mutton, onions, pumpkins and potatoes from his camp oven. When Harry was out of earshot Charlie said, "Sadleir, we knew this bloke could cook, he was station cook on the joint next door, Joe Phelps' run, Windomal, for more than a year but he got sick of it and he came here. Nicholas Chadwick gave him a job as a shepherd. We rigged the draw. Hughie and I can't cook to save our lives and we didn't want the responsibility."

The drive proceeded to McLeod's Crossing using the routine Augustus had taught me on the way from Albury, except Stanley and his boy worked in erecting and dismantling the yards every day, and Charlie and Hughie drove the herd, and as they did so, I rode among the travelling horses and marked those already identified for sale at McLeod's Crossing with a paint pot I carried attached to my saddle. I marked them on the wither, the near or off rump, or the near or off shoulder, according to the prices John Phelps and I had agreed in the yards at Cannally. I described the horses and their prices in a notebook. I was able to mark three-quarters of the sale horses in this way. Those that remained, that I could not get close to in the open, I marked with the help of Stanley and his boy in the yards at night.

Stanley visited his family briefly on the way. I hoped to see Rosie but she was away herding their milking goats, or that's what her mother said. No more pretty looks for me!

We got to McLeod's Crossing in ten days. People rode out to meet us. I don't know how they knew we were coming. Blackfellows must have noticed us and told the townspeople. They badgered me to sell them horses they had selected immediately. I demurred. I felt a responsibility to sell the horses at the best possible price. I told the would-be buyers that I would conduct an auction of the horses for sale within a fortnight and that I would post notices in the town informing them of that. They were disgruntled, but they rode back to town.

Stanley and his boy maintained a camp with a rope yard two miles upriver from the township. They tailed the horses each day for feed and water and cooked for themselves. Charlie, Hughie and I worked near the town – building timber yards for the auction. Harry cooked and shopped for us.

I purchased a light dray from a wheelwright in the town and used it to collect timber for rails and posts along the river. We made a round yard for the sale and two adjoining yards connecting to the round yard with slip rails. In ten days, the yards were ready. I employed a local auctioneer at a commission of 5% of gross sale price, I had notices prepared and pasted all over town and the auction began on a Wednesday morning ten days before Christmas.

I was nervous. I had never been in charge of a commercial affair of this magnitude before, Blue. But I think the nervousness kept me alert. I stood beside the auctioneer with my book of accounts as he started the bidding and immediately knocked down the first lot – a fine Clydesdale mare to a short florid man who had made the first bid for £10. I refused to approve the sale. "Sir, the price is unacceptable. Please put the mare up again for sale." He scowled, but he started the bidding again. The same buyer won the auction again for five shillings more. I intervened again. The auctioneer was called Silverton. "Mr Silverton". I said, please indicate the people who bid for this horse." He glowered at me. "Surely you saw them, Sadleir." There were 50 people present. I raised my voice and spoke to them. "Ladies and gentlemen, I believe that many of you have not had the opportunity to bid on this horse. Perhaps we have run the auction too quickly. I will conduct the auction myself. I will slow the procedure. I will wait for at least 15 seconds repeating the price of the last bid before I close the sale. If you believe I have overlooked your bid, please shout at me to gain my attention. Each horse has a reserve price. If I do not sell it at auction, I will announce its reserve price and it will remain in the yard for you to close the price with me. Mr Silverton is dismissed. My men will mark horses you buy with a paint mark of your choosing. You may not collect your horses until you settle with me. This may take place over the next two days. Have you any questions?

Several people smiled and started to clap, Blue. I learned afterwards that Silverton was a disreputable rogue who was well known for charging an additional commission to buyers of livestock. It was old-fashioned extortion. He tried and failed to persuade several fellows to leave the auction with him. We watched him in silence. Eventually he walked towards the settlement alone.

I had been to horse auctions in Tipperary with my father and older brothers so I knew the form. Buyers were slow with me at first but I waited for them. They were unused to auctions. I learned afterwards this was the first auction run in this way at McLeod's Crossing. It was a huge success. Overall Chadwick and Phelps realised 20% more than the values we had placed on the horses at Cannally. Only two bidders failed to pay for and collect their horses on time so we walked those remaining back out to the herd Stanley and his boy kept in the rope yards out of town and prepared for drive up the Darling.

Before we left I discovered Stanley's *boy* was a girl – his concubine. I was stern, "Stanley, you have deceived me. You must take this girl back to her family at Cannally."

"I cannot Nicholas. She has run away. If she returns she will probably be killed. We love each other. You have seen how well she works on a drive. Please let her stay."

I spoke to Charlie and Hughie. They told me they knew about Stanley's arrangement. They knew the girl from Cannally. "She was supposed to marry an old bloke who had two other wives further down the river. She didn't want to. We reckon Stanley saved her in a way."

I weighed the evidence. In the end I relented. "Stanley I do not want you to interpret this as my approval for what you have done, but you and your boy may stay with the drive to Tarcoola. What is her name?"

"I call her Billy."

We finished the sale's business in McLeod's Crossing after four days. I banked more than £2,000 in coin, banknotes and cheques from the sales to the account of Chadwick and Phelps with an agent of the bank of NSW, and we five continued to Tarcoola. Harry stayed with us as cook. We used the dray I bought to make the yards to carry the cook's store.

We finished the drive at Billilla, a Chadwick and Phelps run near Wilcannia, in late February – nearly three months away from Albury. I felt perhaps I had earned John Phelps and Nicholas Chadwick's confidence. We lost only one horse on the way (it put its foot in a hole, twisted it and broke its front leg, so I had to shoot it). But that was the only calamity. We had good prices for the horses at McLeod's Crossing, the start of a stud breeding troop at Tarcoola and a new lot of working horses at Wurtindeli and Billilla.'

…

6

Managing Albemarle

Paddle steamers travel the Darling and improve life for Europeans living near it. It is the reverse for the Barkindji people. Their food source declines. Nicholas becomes manager of Tarcoola and Albemarle and repays a debt to the guide who led him to Tarcoola. He suffers homesickness. He learns direction of European and Aboriginal people. Nicholas visits Melbourne and Adelaide to appraise markets and suppliers. He invests in a cattle station near Eulo in Queensland. He walks stock to it from Colac in Victoria and ships fencing wire and supplies from Melbourne via rail to Echuca, on paddle steamers to Wilcannia and by camel strings to Bingara.

Nicholas continued with the story of the horse drive after the auction in Wentworth: 'Nicholas Chadwick joined the drive north along the river at Tarcoola to do the cooking for me, Charlie and Hughie. Harry stayed at Tarcoola to work as a shepherd. Ah Chee remained as cook at Tarcoola to look after Stanley and Billy and the shepherds. It was good to have Nicholas Chadwick's company and he told me something John Phelps had alluded to but not confirmed – the partnership of Chadwick and Phelps was being dissolved at Cannally and Tarcoola. Chadwick would keep Cannally and John Phelps would retain Tarcoola. Joseph James Phelps (whom I had not yet met) would move further up the Darling to start to develop Wurtindeli but he would keep Windomal beside Cannally as his headquarters. Billilla would continue to be jointly managed.

"As far as you're concerned Nicholas, John Phelps has plans for you. He and I have discussed it. He asked me to raise it with you if the delivery of horses from Albury was successful. When we offered you the job of head stockman at Tarcoola last year we had decided to dissolve our partnership, but John proposed, because he was a distant relative, that he retain the responsibility for employing you. What that means practically is that he pays your salary. We still have a partnership in Billilla and so from time to time you will still be doing things for me, and I welcome that."

I was somewhat nonplussed, "Mr Chadwick, who owns these horses we are driving?"

"For the time being the troop still belongs to Chadwick and Phelps. John, Joe, and I are still having an accounting. It is entirely amicable, Nicholas, it may take more than a year to sort out the ownership of horses, sheep and cattle and where they stand."

"Do you keep proper set of accounts, Chadwick?"

"It depends what you mean by proper. We do write things down from time to time, but mostly we just divide costs and returns as they occur. We have employed bookkeepers, but it is difficult to keep them. In past years, many of them went to the goldfields so John and I relied on our own notebooks. Are you a bookkeeper as well as a horse breaker, Sadleir?"

"Well no, not entirely, but I have experience of a similar arrangement with partners on the goldfields. In the last partnership I had, I was the only one who could read and write properly so I set up rudimentary accounting systems. I think it helped with the good order of things. We avoided disputes."

"I agree that it is time for something more mannerly. As I explained, John and I just divided things, but we have both realised that our partnership has not been simple. Individually we are responsible for rents on different runs, and the size of the runs and the amount of the rents are not the same, and we have not equally ascribed the value of the grazing to our jointly owned flocks, herds and mobs, and so on. That is the main reason for the division. We both want to be more scientific in our own way. John wants to breed fine horses, and I'm interested in starting a sheep stud to improve our wool. We can continue to assist each other commercially by selling improved breeding animals to each other. The whole thing will be more businesslike. And we need to think about the differences in our families. I have no wife or children but I have sisters and nieces and nephews."

We pushed on up the Darling. It was a beautiful drive. The river was low, but there were plenty of waterholes. We had fewer horses to look after as we progressed. If there was a drawback, it was the heat and flies. They drove one mad, crawling into eyes and nostrils. In my whole time in the backcountry, I never really got used to them. Nor did I find a successful way of combating them. I just kept waving them away with my left hand.

Unlike the flies, I soon became accustomed to the heat, Blue. It was a dry heat – not like the heat of the tropics or the heat of the doldrums on the way out on the Cambridge. Clothing never got wet with perspiration so it was never particularly uncomfortable unless you were working very hard, but – and this was important – everyone needed to drink plenty of water. Several of my people became very sick just by forgetting to drink water regularly when they were working hard in hot weather, and of course, if they were away from water they could be dead within a day. I would not allow people to leave homesteads without carrying water at any time of the year if they were moving away from the river.

We got to Wurtindeli in about 10 days. Feed was scarce along the way, so we made it a slow trip, allowing the horses to feed as much as they could.

Joseph James Phelps stood in front of a tent when we arrived. He had started to establish a station at Wurtindeli. He was welcoming, he knew my function, but he was busy taking stores out to his shepherds. He looked forward to me handling his horses, but he had no men to spare to help me and so he suggested that I return "in a few months". He spent no more than a day with us selecting the horses he wanted, he assigned them to a horse tailer, and he waved us off up the Darling to Billilla.

Hector Robinson was in charge at Billilla. He had a slab hut as the station store and headquarters and he used bough sheds for sleeping, dining and general living. In many ways, it copied Tarcoola, it was beside the Darling, on the same side with glorious red gums in the background, but it seemed even less permanent than Tarcoola and it was considerably less tidy. Robinson seemed harassed. He was defiant and apologetic at the same time. "Chadwick, I was expecting horses sometime as you know, and thank you for them, and thank you for the small load of stores you have brought me, but I am blessed if I know how I am going to keep these horses with me. My boys disappeared down river this morning, I have only one shepherd who will stay – and I really need three. I apologise for the camp being such a mess, but the cook went down river last week as well."

He smiled ruefully. "Perhaps I should think of you as a rescue mission."

"Yes," Nicholas Chadwick said. "Perhaps you should." And he remained silent with a thoughtful look.

Nicholas Chadwick took charge. Charlie and Hughie tailed the horses and Hector Robinson and I strung a temporary yard of rope about 400 yards beyond the homestead to confine the horses overnight. Chadwick set about tidying the camp. He seemed as meticulous as our housekeepers at home in Tipperary except that he was a gentleman pastoral leaseholder, using a broom fabricated from native brush on a dirt floor in an open sided bough shed in the middle of an Australian wilderness more than 200 miles from a township. Gentlemen adapted to their surroundings – I remembered this special lesson.

Unfortunately for Hector Robinson's permanent position as supervisor of Billilla, Nicholas Chadwick wasn't just a saviour. He found a dozen empty brandy bottles underneath one of the beds in the sleeping quarters. He confronted Hector. "Is this your trouble Robinson? Have you run out of brandy?" And then to my astonishment, Nicholas Chadwick went to his horse, unbuckled the saddlebag, drew out a full bottle of brandy, and passed it to Hector. "This is all I have Hector. Ration it. Treat it as medicine. Perhaps it will get you back on your feet."

Hector walked towards the river with his bottle of brandy. He was about 40 with balding wispy hair, spare, skinny and bow-legged. He walked with his head down. He looked ashamed.

Mr Chadwick and I walked towards the horse yards to help Charlie and Hughie yard the horses and see if the yards needed final adjustment. Chadwick said. "I guess you can discern the problem Nicholas. Our man here isn't up to the job at present. Unfortunately, he looks at most of his problems through the bottom of brandy bottle. Goodness knows how he got it here, but I think I know, I wondered why he had so many packhorses when he left from Tarcoola with about 3000 ewes and a few horses just before you arrived. He took a travelling shanty with him."

"What will happen to him Mr Chadwick?"

"I blame myself to some extent, Nicholas. He is a man who needs people. Back at Cannally and Tarcoola he was not a drunkard, but a pleasant steady fellow. I think he got into trouble up here when he had to make his own way. He probably had an argument with one of the shepherds about keeping sheep out of the folds and on feed for sufficient time for them to fatten, left the argument unresolved, came home, drank too much, abused the cook, dismissed him and woke up in the morning to survey the wreck he had created, and had some more brandy to console himself. By the time he sobered up most staff had left.

We need to take him back to people, Nicholas, but we may not do that until we get some order here. I am inclined to offer Hughie and Charlie increases in wages to stay here, Hughie to look after the horses and cook, and Charlie to look after the sheep and the shepherds. I will speak to them presently. If they agree, and I think they will – I have already noticed the beginnings of something more than ordinary friendship with some of the female blacks – I will ask you to approach Hughie to seek his advice about how you may assist him in handling and educating the horses during the next fortnight with a view to having a useful plant of horses at Billilla before you leave."

"Leave for where, Mr Chadwick?"

"Cannally, Nicholas. We may go overland almost directly south away from the river. There will be good feed and fresh lakes and waterholes. Hector, you and I should be able to make an easy 30 miles a day and still keep our horses well fed and watered. We should arrive at Cannally inside a week. You should enjoy it Nicholas. You will see another style of country."

"Will it not be dangerous at this time of the year, Mr Chadwick?"

"It is a prudent question, Nicholas. In many years it would be. But two years ago we had good floods and the lakes and waterholes will still have good water – and rainfall last spring was good for feed.

We should have a good safe trip. Even so, we will carry water and take a blackfellow guide. We will even carry a fowling piece to shoot for the pot. I expect it to be a pleasant holiday tour.

Charlie and Hughie agreed to stay at Billilla. Hughie grinned at me ruefully when we started talking about a program for educating the horses. "The horse work is fine, Sadleir, but cooking!"

"Perhaps you are paying penance for the deceitful way you treated Harry when we left Cannally when you rigged the draw."

"Yes but Harry could cook." And we grinned at each other.

By the time Nicholas Chadwick, Hector Robinson and I headed south on our *holiday* away from the river and south to Cannally, Hughie had six light hacks he could catch easily that had been ridden, and responded acceptably to bit and heel pressure, and two heavy horses that worked steadfastly in harness. He was confidently educating ten other horses, and he had started work cutting and dragging posts and rails from the river with the heavy horses we had broken to build permanent horse yards. He seemed pleased to stay. He was getting more than his fair share of help with cooking and lots of backward looking sloe-eyed looks from one of the girls from the blacks camp by the river camp he had recruited.'

'Were you never tempted, Great Grandfather?'

'What impertinence, Blue!'

'Hardly, Holas. I am 71 now – three years your senior when you died.'

'Yes. Well I suppose you are sensible, but nobody has ever asked me that so directly before.'

There was a long silence.

'Of course I was tempted, Blue. I was a man. I was lonely. I could see other fellows being comforted by dark lasses. Perhaps it was shyness, but I never approached black women and they never approached me. I have never thought about it before but I think it was something to do with the role I was teaching myself to accept – that of a gentleman pastoralist who needed to keep a friendly distance from people he superintended. I think I imagined Aboriginal people as corresponding to the way I thought about ordinary Irish tenants or labourers at home. They were charming, often talented, poorly educated, ordinary people. They were not of our class. They were not our kind of people. One certainly did not marry them.'

'I fear you are avoiding the question, Holas. I have not mentioned marriage. Have I any great-aunts or great-uncles who were born along the Darling River?'

'I think we have spent enough time on this matter, Blue. We need to maintain balance and proportion in our conversations. I was about to tell you the story of my first overland trip – away from the rivers – the first trip south from Billilla to Cannally.'

I listened. Prurient curiosity would have to wait.

'We had not been able to persuade a blackfellow to guide us to Cannally. Everyone we asked had an excuse not to go with us. I realised long afterwards that we are asking these local Barkindji people to guide us into country that wasn't theirs to guide us in. Their protocols forbade it. Hector, however seemed pleased. "I know the way, Chadwick. Bill Black did the same trip in reverse from Cannally about three years ago with a mob of cattle. He talked to me about it. I know where the waters are. We can travel comfortably without a guide."

Nicholas Chadwick surprised me again. He agreed. I was flabbergasted and perhaps a little apprehensive. Nicholas Chadwick noticed my disquiet. When we were alone he said, "Hector is a good bushman, Nicholas, and I need to help him restore his self-respect. I too know the way but Hector doesn't know that and I do not want him to know it. Thus, you are not in any danger. It is, however, important that Hector lead us to Cannally. It will restore his self-respect. He is a good man whom I normally esteem but without his own respect he is of no use to me or himself."

"What about hostile blacks?"

"There has been no news of trouble for quite some time. The blacks out there remain independent and live in the country as they always have. But trouble is always a risk, Nicholas. I have never had to fight my way out of it. I keep my eyes open and behave courteously, but if we really feared unheralded attack we should not be here." I followed meekly when we set out the following morning with three riding horses, a packhorse and four spare horses to be used in rotation, but my Colt revolver was within easy reach, loaded, in my saddlebag. I told no one.

The first two days out from Billilla were good. It was hot, but not dangerously so, and there was enough breeze to keep the flies away. We walked our horses gently across treeless plains. There was grass, blue bush and saltbush and occasionally patches of mulga and mallee. I missed the river's big eucalyptus. The first night we camped away from water, but we did not use the water we carried for the horses, we just used enough for us – for cooking and drinking; Hector was certain we would reach a freshwater lake by lunchtime on the following day, so we tethered the horses (with hobbles they would have strayed too far looking for water).

Hector's prediction was accurate. On the second day we camped for lunch beside a lake with fresh water and the horses drank their fill. We pushed on to camp that night just before dark at a beautiful waterhole fringed with red gums. We were able to hobble the horses knowing that they would not stray far, and there was lots of feed for them. To make matters more idyllic, Nicholas Chadwick had bagged two Mallee fowl with his fowling piece along the way, and I persuaded the others to let me experiment with the cooking method the Barkindji taught me when I walked from McLeod's Crossing to Tarcoola. I found plenty of wet clay and I wrapped the Mallee fowl in it without removing the guts or feathers. I expect the clay bundles measured about the size of a rugby ball as I buried then in the embers of our fire. The other two looked sceptical. Hector said, "This looks like a slothful way of preparing a luxury, Nicholas. Mallee fowl is a special delicacy. Have you cooked it like this before?"

"No", I replied "but I've seen duck done like this and it was wonderful."

"Where did you see it?"

"On the Darling, about halfway between McLeod's Crossing and Tarcoola. A band of Barkindji cooked about 50 duck in this way and they tasted delicious."

"What were you doing feasting on duck with the Barkindji?"

"I was walking from McLeod's Crossing to Tarcoola with Sammy."

"What! You mean Sammy the blackfellow from Wurtindeli way?"

"Well yes. I think that's where he comes from."

"He's a cheeky bugger. He moves around a lot and he stirs up trouble with the local blacks wherever he goes."

"He was good to me."

"I can't believe it."

So I told him the story. Nicholas Chadwick looked on benignly as Hector listened open mouthed. (That was good because he forgot about objecting to the way I was cooking the Mallee fowl). "I would have been scared out of my wits, Nicholas," Hector said when I finished the story.

"I was too when we left together from the police camp, but somehow he knew and he gentled me – just like you do with a green horse. We became close friends. He even taught me to swim."

"Can you swim? Will you teach me?"

And so Hector I stripped for swimming lessons. He learnt to float without fear and within half an hour he was splashing along. Nicholas Chadwick watched us with a bright smile and cooked a wonderful damper in the camp oven, and so we followed the refreshing swim with a grand dinner. The Mallee fowl was superb. It turned out just as I hoped it would.

We thought we were having a wonderful holiday until it started raining the following day. It poured for a day and a half and a night. There was thunder and lightning nearly all the time, the horses were terrified and everything was sopping wet. It was too wet to sleep or camp comfortably so we just stayed in the saddle tailing the horses and trying to calm them around the waterhole until the rain stopped, hoping lightning would not strike us or the horses. When it did stop raining we camped for a day and dried everything out. A lot of the flour was a gooey mess but we saved enough to keep us in damper until we got to Cannally.

That made us two days late, but I was pleased I'd done the trip. If we had gone around via the rivers it would have taken weeks. It was my first long trip in this country away from the rivers' security. The shortest distance between two points is in a straight line, but you need to be sure that there is water in that straight line! I had a lot more to learn about the country away from the rivers, Blue.

I had been in this country about six months. It felt beautiful and terrible, safe and dangerous but I was being nurtured by countrymen who had endured it and prospered. They, like me, had come from the gentle greenery of Erin where seasons came and went predictably and rivers always flowed. They told me of terrible droughts where most of the livestock died because of lack of feed and water, of drowning stock when floods spread miles from the river, and of the men they knew who died from lack of water or isolation from medical attention. And I could see the fat and thriving livestock all about me but they said, "Have a care Nicholas, it is usually much worse than this."

I was in a quandary when I wrote home to Helena to tell her about my adventures and prospects.

Cannally
Balranald district
New South Wales
10th of March 1858

My darling sister Helena,

I am addressing this to you at Kilkenny Castle in the belief that you continue to prosper there in caring for and educating the family of the Earls of Ormond in comfortable quarters beside a gentle flowing river on the edge of a pretty town at the end of a mild winter and with spring approaching.

I will write of the contrasts in our situations. It is the end of summer here, but far from mild. It is hot, dusty and dry. I too am beside a river but it is not flowing. It arranges itself as a series of muddy ponds. It may flow again in spring. People tell me it is fed from melting snow in the Australian Alps, but we are far from the Alps and the supply of water is not always reliable. Sometimes there is not enough snow for waters to reach here. As for my quarters, I usually sleep with a couple of blankets in the open air on the ground (and when it is hot, as it is now, with no blankets at all). It seldom rains here. When it does, and if there is housing available, I sleep in a bed in a house with no walls and a roof made of the leaves of trees.

I guess you disbelieve me. Truly, my dear Helena, these are my real circumstances.

I think you have news of my abandoning gold mining, and I hope you have received a modest share from it from a letter of credit I sent to Pater some months ago. Richard and John were pleased with my leaving the goldfields (and they continue to behave like our parents in urging me to pursue a legal career in Melbourne) but instead I wrote to John Phelps, mater's cousin from Limerick, who has been long established in New South Wales, to seek his advice about a calling in animal husbandry. He invited me to join him at one of his sheep stations along these rivers last September and since then life has been a series of adventure in learning to live and prosper here.

I am now at my wits end. This country is beautiful and dangerous but I fear I will never understand it. I want to persist and seek my fortune as our cousins have but I feel daunted by the challenges of raising livestock profitably in such a variable environment. It is good to write to you Helena my dear sister. Doing so has relieved some of my present melancholia.

I have been called to luncheon – a repast of salt beef, unleavened bread, and boiled onions.

I have just reread what I have written and feel inclined to cancel it because I feel optimistic and cheerful now (perhaps it was the boiled onions), but I will leave what I have written to demonstrate your confidential importance to me.

I trust things continue to go well with you. When I have made my fortune, I will look forward to welcoming you to this beautiful land of drought, flies and flood.

Please write to me when you can. Mail is slow here, but I treasure letters from home.

Your loving brother.

Nicholas Sadleir.

John Phelps had remained at Cannally, drafting livestock to help with the division of assets of the partnership with him and Nicholas Chadwick. He told me that by the end of the week he would have a flock of rams for me to take up river to Tarcoola, Wurtindeli and Billilla. He wanted me to deliver about 200 rams at each station, alert the overseers and shepherds of the need to remove the old rams and replace them with new ones in all of the flocks and arrange for a flock of the old rams to be run separately. He wanted me to count all sheep, cattle and horses and to report on their class and condition. He asked me to gather all aged saleable sheep, including the old culled rams, and bring them to Cannally.

We talked of the horse sale at McLeod's Crossing. I gave him my book of accounts and the receipt from the banker's agent. We talked in his office and he called the bookkeeper from a smaller room beside his to transfer my records to his.

"This was a grand affair, Nicholas. Congratulations. How do you achieve such a good result?"

I no longer blushed when John Phelps complimented me, but I still knew he meant it. "We built horse yards and ran an auction outside the town. I employed Mr. Silverton, a local agent to conduct the sale but I observed very early in proceedings he was cheating us by selling cheaply to a conspirator. I dismissed him and conducted the auction myself."

John Phelps looked up from the figures on his desk. "Really? Nicholas! You? Like the auction touts at the horse sales at Cork and Limerick?'

"Yes," I grinned. "I did not have their line of patter, so I behaved like a dull gentleman. I spoke slowly and waited for their bids. It was something new for us all. The townspeople told me that there had never been an auction at McLeod's Crossing, and certainly, I had never been an auctioneer."

"And the yards are still at McLeod's Crossing?"

"Yes we left them intact. Perhaps some of the townspeople are using them. I am unsure about who leases the land we built them on."

"That is of no importance at this stage. I believe the government of New South Wales is conducting a survey to establish an orderly township at McLeod's Crossing. We should assert our ownership of the yards. We may wish to conduct further sales. Can you do that on your way to our runs on the Darling?"

"What if I mark them with our horse and cattle brand."

John smiled. "You are coming to rude colonial habits quickly Nicholas. That should serve. And could you make a note of record at the police camp?"

John went on to explain his reasons for sending new rams out to the stations along the Darling. "Nicholas Chadwick has applied his learning to the improvement of our flocks by establishing a ram breeding stud here at Cannally. He has pure Rambouillet rams and has built up a stud breeding flock of more than 2000 ewes. These ewes are already cutting an average of a pound more wool than our ordinary flocks, the wool is finer and the price is better. It will take about two years for these new rams to make any difference to our wool clip, but when they do, it should improve our returns per head."

"How do you compare returns to your costs, John?"

"That is a fine question Nicholas. Wool prices are not stable. If anything they are diminishing, but if they were not, and if we got no drought within three years we could profit by, say, four pence per head. Calculate that for 100,000 sheep."

"Nearly £1,700."

"And it is money for little additional effort. We need no more shepherds or shearers, no additional rations, just the satisfaction of knowing we are breeding superior animals."

"And have you an estimate for the numbers of additional sheep and cattle you may have on the runs you have now, John?"

"We have not experimented with improved grasses, and I doubt we can. I think in this country we should rely on what grows naturally, but if we account for a natural balance of seasons – say a large flood every five years, average or below average seasons in four years out of six, and a drought every seventh year – with good overseers and good shepherds I can imagine doubling the numbers of sheep and cattle we have now. We are using only a tiny proportion of the grass we grow. At present, our limitation is competent shepherds. We need people back from the goldfields."

"Will you buy stock or breed them, John?"

"I think we can breed them without risking additional investment, Nicholas, and we can preserve the quality of our livestock with the selective breeding we do. With our sheep, for example, we could double our numbers of breeding stock just by natural increase within four years, but I think we should go more slowly. We should class our sheep following Nicholas Chadwick's example, and reject up to 30% of our maiden ewes, remove them for sale, and use rams only from a stud. And, I think we should continue improving our flock by getting rid of our older breeding ewes before they are eight years old. That helps to improve the flock as well – the improved young continually replace the inferior old and we have breeding stock or fat stock to sell. I feel that Nicholas Chadwick may not be able to supply all the rams we need from here at Cannally, so we should look to establishing our own ram breeding flock. Perhaps you can think on that Nicholas and give me some more advice when you have returned from the runs with the sheep for sale. And please consult Joe when you see him at Wurtindeli. He has some firm views on the matter."

"How many men should I take with me to drive 600 rams?"

"They are a small number, but you will have to run a night watch, so I think you need three extra people and a cook. You should be able to hobble your horses at night and share horse tailing duties during the day. I invite you to speak to Tom Jacobs our overseer, you know Tom?"

I nodded.

"Tom will nominate the people he can spare. Ask for eight people and choose the four you will take from those. Tom may propose people he wants to get rid of. Tom will see you right for horses as well. Make sure you get a good pick."

"And rations?"

"Carry enough only to get you to Tarcoola. All stations should be well stocked and you can replenish your tucker box as you go. Ask Tom Jacobs for 15 wethers to go with the ram flock so you can use them for rations. Things are getting dry. I doubt that you will find much meat to shoot worth eating along the rivers just now."

I spoke to Tom Jacobs about needing eight men for the drive. Tom was tall, paunchy and broad-shouldered and he leaned back against the rail of the horse yard and grinned at me. "Nicholas, you don't need eight. You need no more than four. John Phelps put you up to this Nicholas – picking the eyes out of the people I nominate and taking only half of them. He does it to me every time we move stock away from Cannally. I reckon he knows that I will stay here looking after things for Chadwick and I'm pushing all the undesirables towards him for the stations along the Darling.

"Well perhaps he is right, Tom."

"Nicholas, this business of picking up people on the station and in the droving camp isn't like classing rams in a sheep yard. If I run eight blokes past you and you choose four, what do you think the four you left behind are going to think of me? Or for that matter, even the ones you take?"

"What do you suggest, Tom?"

"Well I reckon they need to be volunteers, Nicholas."

"And you will simply go to four blokes you don't want and say you, you, you and you."

We stared at each other with our arms folded.

I said. "Tom, what if I go to the men's dining room tonight, tell them about the drive and the trip back and ask for volunteers. I can tell them I will be the captain and that they probably have a great deal to teach me. They may like that."

"And what if you get no one?"

"Then you'll know that every station hand and shepherd on Cannally loves you, Tom."

"And what if everyone puts up their hand?"

"Then you'll know that everyone hates you, Tom, and that you should come with me, volunteer to be the cook and pick three others as assistants."

"Are you serious?'

"Not about taking you as a cook, but I am serious about asking for volunteers tonight."

"Well you're game."

"Let's see how it goes, and now, let me have a pick of the horses."

We walked to the larger yard.

"Volunteers?"

"Absolutely. Watch them run to me through the gate." And we rolled into the yard bursting with laughter.

We had a wonderful time chasing each other and horses around the yard –Tom pushing inferior animals towards me at the gate I was trying to control, me pushing them away and at the same time trying to coax some good horses through the gate. Tom gave up in the end. He was puffed out. He said. "I give up, Sadleir. Here, take these."

And he pushed up the eight horses I had been trying to persuade through the gate. He knew his horses.

We walked out of the yards. "And now," Tom said, "for the persuasive eloquence and measured debate of the men's dining room."

I think Tom was disappointed. The men were eating mutton stew hungrily when I walked into the dining room. I banged a tin plate for silence and said. "I am Nicholas Sadleir, many of you I know and some of you I don't. I am seeking three assistants to walk 600 rams up the Darling River, 200 each for Tarcoola, Wurtindeli and Billilla. It will be a round trip. We will be collecting sale sheep from each of those stations and their runs for assembly here for a drive to Melbourne. We expect to be away for three months. We will pay you a bonus of two shillings a day for every day you're away. Please speak to me at breakfast time tomorrow if you're interested." And I walked out.

Tom chased me. "Nicholas, that was a bloody bribe."

"To preserve your feelings, Tom. You will never know whether the people who volunteered did it because they detest you, covet additional money or esteem me. And," I paused. "You may not come."

He laughed and swore at me again as he stumped off to his quarters. "You take the cake Nicholas. Cannally will soon have nobody left."

The following morning, before breakfast I confessed to John Phelps. "I am afraid I felt obliged to offer a 10% wage rise to persuade people to come with me to Billilla. But I asked for three only – not four so the overall costs will be less. I hope you approve."
He allowed himself a slow smile when I told him about the games Tom and I had played. "You may continue to make these small commercial judgements without consulting me or Joe, Nicholas, but please make sure you keep a record and that you transfer your records to the partnership books as soon as you can. And please discuss it with Joe at Wurtindeli."

He continued. "I notice more people are coming into the district looking for work from the goldfields. This is an excellent sign for us. We need more people. If you see good coves, Nicholas, recruit them as general hands at about 16 shillings a week, and fully found, [1] and put them with our experienced men as probationers to do whatever they are doing."

'It was much like that for the next four years, Blue. I was moving stock backwards and forwards up and down the Darling as we started a grand plan to use all of the country we leased – so far we had concentrated our flocks close to the rivers. Nicholas Chadwick was establishing his independence; John and Joe were working to a formal partnership between themselves, because until then they had each leased country in their own names. Nicholas Chadwick got Cannally and Billilla and he started Kallara further up river. John and Joe Phelps kept Tarcoola and extended the Wurtindeli runs further east to include Victoria Lake and they called the new station Albemarle – after the Duke of Albemarle, a distant ancestor.

John and Joe Phelps and Nicholas Chadwick and I talked a lot in those early years. We dreamed about ways of using the grazing away from the river all the time – not just in winter when the dew on the grass and the moisture in the green feed gave the sheep enough water without flocks having to go to the river. We knew floods every five years or so left enough sheep feed for years many miles from the river but we couldn't use it for most of the year because there was not enough stock water out there. There were large freshwater lakes but they needed a huge flood to fill them and we wanted more water than they could provide, and in other places.

"We need to sink wells."

"We need to set up shepherds' huts each with its own well, windmill and water trough."

"Perhaps we can divert water from the river further out using dams and channels."

And when paddle steamers first went up the Darling in 1859 we delighted in the possibilities of carting wool from a centralised shearing shed on the banks of the river.

"We need one large shed for each station."

"We need to walk the sheep in for shearing. In that way we arrange for them to freight the wool on their backs to the river."

---

1. Fully found means all food and lodging provided

"We will need to abandon shearing out on the runs at each of the huts."

"The large sheds on the banks of the river can be designed like mill factories."

"We can set up better accommodation and cooking and dining facilities for shearing staff."

"If men get sick or hurt they will be closer to help."

John, Joe Phelps and Nicholas Chadwick travelled frequently. They returned from trips to Melbourne and Sydney with the news of experiments in sheep and cattle husbandry in other parts of the colonies.

"People in New England are crossing Merinos with British breeds (mainly Lincolns) to improve their size and lambing percentages."

"Perhaps we could try a small experimental flock."

"Yes but we should isolate the flock in case the experiment fails. The pure Merino strains we have developed need to be preserved."

And most drastically of all, we evaluated the radical idea of containing sheep, cattle and horses in large paddocks without people in attendance and using wire fences instead to keep them from straying. Graziers close to Sydney were starting to do it.

"They will become wild animals."

"What about the care of ewes during lambing? Lambing percentages will be terrible without shepherds."

"It will be a wild dogs' banquet."

"The costs of fencing will be prohibitive."

"But shepherds and their dogs cost a lot too."

'The paddle steamers transformed the river. Sam Curtis and others were at the forefront. They gave us a fast reliable service. The wool went away from the riverbank and we were able to order stores cheaply direct from suppliers in Goolwa near the mouth of the Murray in South Australia and so we were able to keep large stocks of food and clothing all year round. It was a convenience to those who worked on the stations. Everyone there could have an account at the station store held as a balance against his or her wages. Tarcoola and Albemarle became small towns. And apart from allowing us to develop good station stores, the paddle steamers themselves became shops. Many carried things station stores didn't have – fine dresses for the ladies, books and magazines, sewing kits, millinery, all sorts of haberdashery and jewellery. Later still, a few paddle steamers became floating churches. There were regular church services at all the stations along the river we had several weddings on steamers at Albemarle and Tarcoola over the years.'

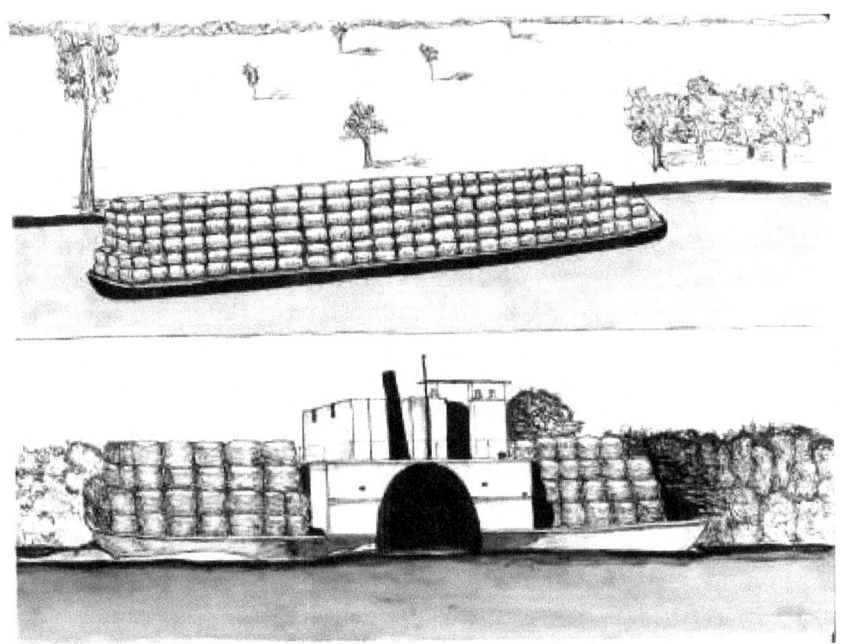

Barge and Steamerr loaded with Wool

'What happened when the Darling went dry?'

'Steamers would rest on the bottom where they stopped for lack of navigable water. Most of the crew would leave to walk back to where they came from, and if there was valuable cargo perhaps one or two people would stay with the boat and camp there to keep it company. Usually it was only for a month or two before water would come rolling down the river from rain and floods in Queensland or eastern New South Wales. It was only for a couple of years that wool was in the shed at Albemarle for more than 12 months before a steamer was able to collect it. Two years was the longest I can remember boats stuck in the Darling.

And interestingly, the cheap freight on paddle steamers persuaded us to stop washing sheep before we shore them – or at least that was part of the reason – we had trouble employing people to do it as well, but we had justified it by thinking about the money we were saving in not carrying the extra dirt in wool on expensive bullock wagons.

We got horse feed up on paddle steamers too – grain and hay and chaff. It was costly but wonderful to have when you were in a hurry on a long journey. And inevitably, as a result of that we thought of establishing farms to grow horse feed in good years.'

'Did you try it Holas?'

'Yes we did. We had some success, but they were mainly bonuses. If we sowed crops after a flood we knew we would get a good hay crop, but that was in a year when we didn't need additional feed anyway. In years when we didn't have a flood and sowed crops on the first rain, the crops failed more often than they thrived. In the end we abandoned farming and had chaff sent from Morgan or Echuca.'

'But we are getting ahead of ourselves, Blue. We didn't try farming until much later. In the early 1860s, Nicholas Chadwick and the Phelps brothers were still teaching me the country. We had two dry years, one drought and two good years, but we still managed to increase our flocks because we were able to use more country as we sank wells.

I was living in stock camps and droving camps for most of the time but John Phelps gave me a holiday in Melbourne to attend brother Richard's wedding. Sam Curtis carried me from the Tarcoola wool shed to McLeod's Crossing. He left me there as he turned down river to Goolwa with a load of wool and I was able to join another steamer going up the Murray to Swan Hill. I took a stagecoach from Swan Hill to Sandhurst (it was called Bendigo by then) and took my first steam train journey to Melbourne. That was a thrill. I was one of the first passengers on the new line.

Melbourne was magnificent. Its streets were paved and it seemed more sober and respectable. I'd not been there for four years. There were still gold in mines in the colony, but it wasn't in the hands of independent diggers offering brandy to all comers on street corners. Respectable mining companies had it, and Melbourne exuded stately probity. There were new churches everywhere.

Richard married Eliza Grattan at a church in Toorak. I was pleased to see John there as well. His wife Isabella had not been able to come. She was at home at Hamilton looking after their three children.

We talked of our lives and news from home. Richard had a letter from Brookville House telling us of a railway installation between Cork city and Bandon on the south coast. Marshal had ridden on the new train on a holiday trip to Cork and Kinsale. Pater suggested peace was returning to Ireland because of the number of trial hearings for murder and violence were diminishing. We grinned. Pater was always more interested in legal statistics than we were.

Richard's hospital was successful. He was making a good living and he had a good reputation. Eliza, his new wife, looked forward to assisting in the management of the hospital and they hoped to expand it in a year or two.

John delighted us with his police stories. We talked of Robert O'Hara Burke who had recently passed through my district on the Darling on the way to the Gulf of Carpentaria and to his subsequent death from starvation on the way home.[2] I had missed the expedition. I was at Billilla breaking horses, but I'd heard all the news of the party and its extravagances as they moved up the river. Robert O'Hara Burke was John's superintendent at Beechworth when Robert was appointed to take charge of the expedition.

"I wanted to go with him." John said. "He said, *You may not come John. You have children.*"

John regretted not being able to go. His stories of Burke's eccentricities were hilarious and I think John felt that he might have been able to modify some of Burke's extreme judgements on the expedition. John seemed sombre. He had lost a friend who had taken a role for which he was unsuited and had perished with his colleague as a result.

---

2. Robert O'Hara Burke led the ill-fated "Burke and Wills" expedition from Melbourne to the Gulf of Carpentaria. He and William John Wills died of starvation in the Cooper's Creek region on their way home in 1861.

I basked in the luxury of Melbourne for a further three days and took a steamer to Adelaide in the colony of South Australia and endured another bout of seasickness. I stayed in Adelaide and found it an agreeable and elegant city. I called on several business houses there. John Phelps had asked me to explore possibilities of improved commerce with South Australian traders and shipping agents now our wool went down the river to Goolwa, by rail to Victor Harbor, and thence by clipper to the wool sales in London. I visited Port Adelaide (little did I realise that I was to establish an Adelaide home at the Grange for Madam and 12 children not far from the port – but that was not until years later). After a week, I travelled to Goolwa by stagecoach skirting the agreeable Adelaide Hills. It was a bustling port with goods going to and fro. I greeted many boatmen I knew from their work along the Darling. I learned that Sam Curtis had left to go up river two days before. I was sorry I had missed him but there were plenty of other boats leaving for destinations along the river and I was able to secure passage easily. I found a comfortable steamer towing a barge to collect wool near Bourke on the Darling so I was pleased to join her. She had cargo for several merchants in McLeod's Crossing, but its unloading there was the only likely cause for delay, so I expected to be at Tarcoola in about a week. It seemed a strange way to have a holiday – moving along a waterway where I spent my life trailing sheep, horses and cattle, but I was not obliged to do that and I sat and read books from a replenished library bought in Melbourne and Adelaide and watched magnificent red gums sliding by. I never tired of my new Australian rivers.

When I got to Tarcoola John and Joe Phelps summoned me. "We are pleased you enjoyed your holiday Nicholas," John said. "Joe and I both seek one. I intend to return to Limerick for a period, and Joe will be away in Sydney. We want to offer you the post of manager of Albemarle and Tarcoola. We want you to establish your headquarters at Albemarle. You have a free hand to appoint and dismiss assistance as you see fit and to manage all systems of accountancy and bookkeeping. Could you please propose improvements to both runs for our consideration before Christmas this year? We have money to invest and we want you to manage it.

We offer you a salary of £300 with all accompanying perquisites. What do you say?"

And so I became manager of Albemarle, and as you know, Blue, it was for the rest of my life.'

Nicholas Clarke Sadleir aged 27 years

'What about the Barkindji Holas? How were things changing them?'

'I hadn't realised when I first arrived on the Darling, Blue, but Aboriginal people in the district were plagued by diseases from Europe. Some of the old hands told me about it. The early explorers noticed families with sores resembling smallpox and people sick and dying. There were also stories of fights, and settlers killed, and massacres in the district, suggesting some blackfellows were murdered, but there was little of that in the time I was on the Darling. Again, people said stations had been abandoned because of blackfellows' resistance, but I never saw it. Perhaps I was too late. What I did notice was that the Barkindji seemed to rely more on the stations for food. When I first came to the district I saw stores of grain harvested from native millet kept in small silos roofed with straw from native grasses. The women carried special grinding stones with them to make flour from the grass seeds. These grain stores disappeared quite early in the piece. I didn't make a study of it; perhaps our sheep, cattle and horses reduced the supply of grain from the native grasses, and the Barkindji relied on us for flour when they could not get enough millet seed to make their own. On Albemarle we never stinted them. Quite a number of families were happy to serve as hut keepers and shepherds away from the river, and as long as we visited them regularly, treated them kindly and provided good rations, they made good workers. And like everyone, they needed praise.

As we built up our station homesteads to become small townships, we employed many of the women as housemaids and cooks. They camped with their families along the river, bathed in the river and changed into dresses we provided before they came to cook and clean. They did good work.

'How did you pay them?'

'We provided the whole family with rations – as much as they could eat, and we paid anyone who was working at the same rate as we paid the white blokes. Joe Phelps insisted on it.'

'How did they collect their wages? What did they spend money on?'

'Well that was the trouble, Blue. We ran accounts for them in the same way as we ran accounts for everyone and we offset their purchases at the station store against their accounts. But they seldom asked for money and they had no idea of accountancy. Some of the people understood they were being paid in goods – clothing, hats, mirrors and so on – and it was rare for somebody to ask for more than they were owed. There were seldom arguments, and if there were, we usually mollified unhappy parties with a gift. The Barkindji people simply didn't think commercially in the way we did. They didn't use money normally. It seemed to me that the station had a duty to protect them but it was hard to persuade storekeepers and bookkeepers to respect that. I think it's sad to say, but we probably exploited them because they didn't think in account balances in the way that we did and their credits disappeared.'

'And what about interbreeding, Holas?'

'What are you getting at, Blue?'

'White station hands, shepherds, hut keepers, shearers, boat men, drovers, wheelwrights, blacksmiths, well sinkers and all the rest, taking aboriginal concubines. What happened to the children of these unions?'

'Oh I see what you mean. We called them half-castes but they mostly lived as blacks. Usually their white fathers didn't stay with the Barkindji women for long. They were mostly casual encounters so the black mothers returned to their family groups with their yellow-skinned babies and the children were raised happily with the support of their clans, but it wasn't always like that. You remember me telling you about Sammy, the blackfellow who took me from McLeod's Crossing police camp to Tarcoola when I first came into the country.'

'Yes.'

'Well a young aboriginal woman called at the station and asked for Nic'las. I was there at the time and I spoke to her. She had a small yellow-skinned daughter and her mother was with her. She told me they had walked down the river from the other side of Billilla. Her English name was Lilly and she told me that she, her mother and her baby daughter had been banished from the clan there because of her father's misbehaviour. Her father was Sammy. He had disappeared, fleeing east with tribal executioners chasing him. Lilly did not expect to see him again and she feared retribution against her mother and her baby and herself. Sammy had told her before he fled to come here and look to me for protection and help.

It was a dilemma for me, Blue. The police at McLeod's Crossing were three days' ride away. The local station blacks made it clear the trio were not welcome with them. They wanted no trouble with blacks from further up the river, but if Sammy's wife, daughter and granddaughter needed protection, then I owed them sanctuary. Sammy had named me specifically. It was a serious debt. I had my sleeping quarters and office in a separate log hut with a double rail fenced yard surrounding it. I invited them to camp in the yard in exchange for keeping my hut and office clean.'

'So you inherited a family, Holas?'

'I suppose I did in a way. Talking to you about it now, Blue, helps me appraise my feelings at the time and how I reacted. I had just received a letter from Richard telling me my father James had died at Brookville House. I still remember how remote and lost I felt – so far removed from the comfort and closeness of Tipperary. Because the mails took so long he had been dead for more than four months, but to me it seemed as if he'd died yesterday. Perhaps taking on another family obligation helped me to adjust. I was too far away to help my mother in her widowhood so perhaps I compensated in some way. I've never considered it in this way before, Blue. The arrangement didn't last long but perhaps I did myself more good than I did to Lilly and her mother and daughter. '

'What happened to them and you? Were there reprisals from the people further up the river?

'No my guests were never harmed. After a couple of months the local Barkindji welcomed them into their river camp, but Lilly's daughter remained a favourite of mine. She was a toddler when her family camped in my yard and she used to come into my office to play with the pens and pencils when I was there. Lilly agreed to my naming her Pamela. When the trio moved to the river, Pamela came to visit me regularly when I was at home. When I had time, I started to teach her to read and write in English. She grew up on Albemarle and worked in the station store until she picked up with one of our drovers and travelled with him on drives south with flocks of sheep from here. She raised three daughters with him in stock camps.'

'Who was her father, Holas?'

'It was never discussed. It was probably somebody from up Billilla way.'

'Are you sure, Nicholas Sadleir?'

'I'm not sure I like your tone, Blue.'

So we changed the subject. Holas talked about fencing the 50 mile boundary to the south – a straight line separating Albemarle's lands with Tolarno's country. 'We had two gangs starting at either end using compasses to guide them towards the middle, and no, we did not meet at a precisely agreed point. We were about 400 yards out. The fence had a dog-leg in the middle.' He chuckled.

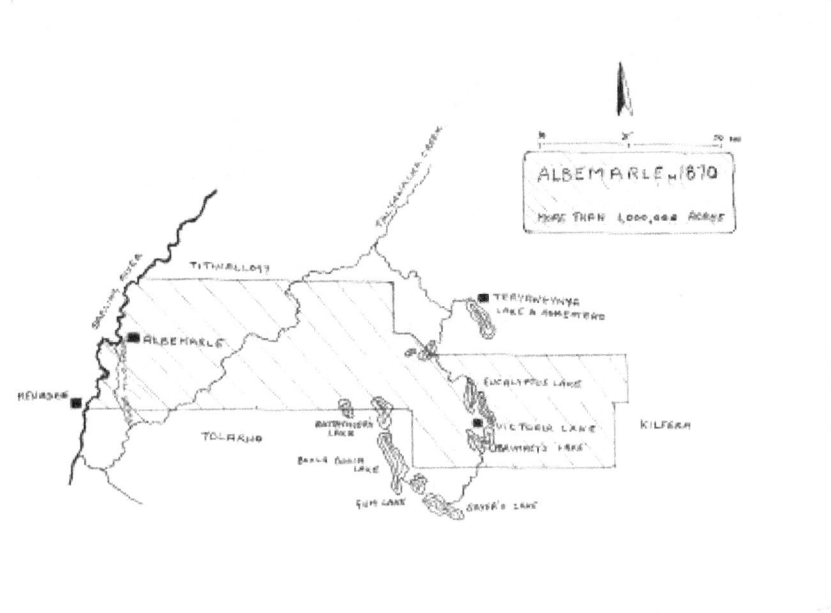

Albermarle in 1870 Copied from Richards – Mousley, Claudia, Big Men Long Shadows: a story of the history and happenings of a sheep station on the River Darling – Windalle, Kensington N.S.W.: Nelen Yubu Publications, 2010 and reproduced with her kind permission.

To change the subject even further, Nicholas told me about a partnership he developed in Queensland with Fitzherbert Brooke near Eulo when he, Nicholas, was managing Albemarle. They called the station Bingara.

I decided to travel there. I was at Hungerford, a tiny settlement 500 miles inland from Brisbane, straddling the New South Wales and Queensland border, sitting outside the pub and talking to two old-timers about how things might have been in Great-Grandfather's day.

'There were cattle camped miles out in all directions, there wasn't a dog fence or a gate here then. There was a bit of a hut beside the fence where the customs collector had a headquarters, but I reckon the local coppers did most of the work. God knows how they sorted it out, I've never heard of drovers with bags of money, but in those days a man's word was his bond. I reckon the owners of the mobs paid by cheque and posted it as soon as they got the bill. That must have been the go.'

'What about the stock routes?'

'Cattle came from hundreds of miles away. Right down from the Queensland gulf country. In the beginning, herds followed rivers but soon governments put in waters for travelling stock – they put in dams, earth tanks, and bores.'

'Were there drovers living in Hungerford?'

'No. They were closer to the source, Cunnamulla, Charleville, Windorah and so on, even further up. This place was a crossroads. They did stop here for a bit of a party from time to time, but only between drives.'

'How do people manage now?'

'Oh everything goes by road transport now. When I first came into this country, I can just remember sheep and cattle being walked to Bourke, but that was pretty well the last of it. Stock go further now too. In the old days just about everything for slaughter went to Tancred's at Bourke. But that meatworks has closed, and there are so many more markets and the station people are much better informed to get the best prices for their stock.'

'A lot of the stations had farms on them.'

'To grow hay for horses?'

'Yair. There were still some old bits of machinery knocking about on the place I worked on when I got up here, and some of the old blokes talked about sowing crops and making hay in good years. They had to. They had to keep their horses going, and on long journeys, in droughts, they carried chaff with them for horse feed. Those old blokes were mighty glad to see the first motor cars. It made life a hell of a lot easier not having to try to grow hay in country where they shouldn't have been trying anyway. Now the fuel comes out on tankers.'

'There's more variety now too. I worked for a bloke with a well-fenced block he ran as a goat joint. Not really a farm or station, more of a depot. People from stations would drop in small truckloads of their feral goats. They couldn't trap big mobs at any one time, just those they could catch, so he bought them as they came. The best goats went out to his buyers that day if he got enough to make up a load. He worked out which goats were big enough to be sent for slaughter and he let the little ones go, but only on his place. He got them back when they were big enough and stuck them on a truck when he had a load. He was laughing.'

'Is it that profitable?'

'Bloody oath. The bastard's making a fortune. He used to send his goats to Bourke, but he had a blue[3] with the bloke there and now I reckon they go to Mildura. The export price is terrific.'

'And even with all the changes, a lot of the big places continue just as they always have. The proportion of sheep and cattle they carry varies with the price of wool and the price of beef. There were terrible droughts recently. Before all this glorious stuff we can see, we had 10 years when we got bugger all[4] rain. Wool prices were terrible, so the old places relied on their cattle to keep the joints going. Even so, just about everyone survived. A few places changed hands but it wasn't forced by drought usually.'

One of the blokes bought a loaf of bread from the publican and they left together walking in the same direction to their separate homes.

'What did you think of that, Holas?'

'It was instructive, Blue. It seems times have not changed much. What did he mean by cattle going to markets by road transport?'

'He meant that the cattle stand on the decks of motor trucks just as they used to ride on railway trucks in your day, and that the cattle are loaded at the station yards rather than having to be walked to a railway siding.'

'How many head can a truck carry?'

---

3. 'Blue' = a fight or argument
4. Bugger all = little or none or nothing

'It depends on the weight and size of the animals and the size of the truck, Holas. Some of them call themselves road trains and they tow the equivalent of two railway trucks, as you would remember them. They could carry about 50 large bullocks or 100 yearlings or 1000 lambs more than 1000 miles in two days.'

'A four-month walking journey in my day, but we would have taken bigger mobs or flocks.'

Nicholas Sadleir fell silent. Eventually he began to reminisce. 'You're at Hungerford now, Blue. Describe it to me.'

'I'm sitting facing the road running north and south on the veranda of the Royal Mail Hotel. It is flat red ground and the whole town comprises no more than 10 buildings. The hotel is made of corrugated galvanised iron and it has bedrooms opening on to the veranda. The bar opens onto the veranda as well – on the north-east corner.'

Hungerford Hotel 2011

'It sounds just as it was when I was last here. I came here first in about 1872 on a stagecoach. I was on my way to meet Fitzherbert

Brooke further up the Paroo to look at some country. I'd met him in Melbourne on a trip down there to the Flemington stock markets. He was looking for partners or capital to develop leases he had, he seemed a fine fellow, I'd heard good reports of country along the Paroo River, I had some cash to invest, and so I agreed to meet him to look at the country. There was a government auction of pastoral leases planned at the time so I came prepared with letters of credit in case I needed them.'

'From the gold you got?'

'Yes, and from other mining shares. Most of the gold money went to Ireland.'

He paused and continued. 'It was a long trip. I left on a paddle steamer from Albemarle for Wilcannia, took a coach from Wilcannia via White Cliffs and Wanaaring to Hungerford and then on to Eulo. I can't remember exactly, but I think it took more than 10 days. I have a jaundiced memory of Hungerford. I was carrying two bottles of brandy as a present for Fitzherbert and I had to pay customs duty on them at the border. I think it was three pence.'

'Someone else had a dim view of Hungerford as well, Holas. I read him a passage Henry Lawson penned in the 1890s.

The country looks as though a great ash heap had been spread out there, and mulga scrub and firewood planted – and neglected. The country looks just as bad for a hundred miles round Hungerford, and beyond that it gets worse – a blasted, barren, wilderness that doesn't even howl. If it howled it would be a relief.

I believe that Burke and Wills found Hungerford, and it's a pity they did; but if I ever stand by the graves of the men who first travelled through the country, when there were neither roads nor stations, water tanks, nor bores, nor pubs, I'll take my hat off. There were brave men in the land in those days.

It is said that the explorers gave the district its name chiefly because of the hunger they found there, which has remained there ever since. I don't know where the ford comes in – there is nothing to ford, except in flood time. Hungerthirst would have been better. The town is supposed to be situated on the banks of the river called the Paroo, but we saw no water there, except what passed for it in a tank. The goats and sheep and dogs and the rest of the population drink there. It is dangerous to take too much of that water in a raw state.

Except in flood- time you couldn't find the bed of the river without the aid of a spirit level and a long straight – edge. There is a custom-house against the fence on the northern side. A pound of tea often cost six shillings on that side, and you can get a common lead pencil for four pence at the rival store across the street in the mother province. Also, a small loaf of sour bread sells for a shilling at the humpy aforementioned. Only about sixty per cent of the sugar will melt.[5]

'It wasn't quite as bad as that when I was there, Blue. Who was the author?

'Henry Lawson.'

---

5. Extracted from 'Hungerford', a short story in a collection entitled *While the Billy Boils*, Henry Lawson, 1896, Angus and Robertson Publishers, Australia

A poster for "The Bulletin"

'The tyke who wrote for *The Bulletin*?'[6]

'Yes.'

'He was an amusing writer, but he always looked on the dark side of things. Sometimes I remember him as being politically dangerous. We kept a small library for the men at Albemarle and his books of short stories and poetry were always popular. I just wished he hadn't been quite so grim.'

Nicholas and I talked again on the following day as I stood

---

6. A weekly journal written and printed in Sydney for outback communities

before Fitzherbert Brooke's grave at the old Bingara homestead. The fine marble headstone announced:

In memory of
Fitzherbert Brooke
of Bingara
born at Avening, Gloucestershire 3rd of September 1838
died 31st of January 1887
"Whosoever shall call on the name of the Lord shall be saved."

I read it aloud.

Nicholas said. 'He was a good man and he died suddenly – possibly of appendicitis – no one really knew. This was on the margins of civilisation then, not like the relative luxury we had on the West Darling. He was too far from medical help. How does the headstone look?'

'It is of fine marble and the lettering is good but there is some staining and fading with age. Did you have it put here, Holas?'

There was no answer. After a silence, he asked me, 'Tell me what else you can see.'

Fitzherberrt Brooke's Grave, Bingara

'There are no buildings. The only thing that is obvious is a broken windmill and a rusty galvanised iron tank, but I have walked around a bit and I can see from the look of the soil where there were buildings and garden beds.'

'Nothing?'

'Yes. The new owners told me no one has lived here for at least 80 years.'

'Yes. I forget about how long ago I was here. This was wonderful country for breeding and we had a good run of seasons. Fitzherbert Brooke and I were partners for nine years. When I got off the coach from Hungerford I was immediately pleased with the potential of the district. Fitzherbert and I spent a week riding around the country with blackfellows and we talked about how we would fence and improve it before we went to the auction in Cunnamulla. I did the bidding. We paid fair prices. We renewed the leases on the country that Fitzherbert already held and we extended them. We called the combined runs Bingara and Werie Ela. We had about 550 square miles, there was plenty of water in holes and soaks along the Paroo River and we thought we could get good water in wells too. We thought we could run 10,000 cattle and a few hundred horses.

'I stayed for about a fortnight after the auction looking at the stock Fitzherbert already had, planning the establishment of the station with him and talked about how we could work together profitably. We agreed that I would purchase breeding animals for Bingara from studs I knew in Victoria and I would set these horses and cattle out with a drover. I was to arrange supplies of fencing wire and other station stores from suppliers in Melbourne, have them railed to Echuca, consigned by paddle steamer to Wilcannia or Bourke and carried by bullock dray or camel string to Bingara. I was well known to merchants and carriers and I could easily establish lines of credit for the firm Sadleir and Brooke. Fitzherbert would manage the station, its improvements and fencing, I would maintain supplies with my contacts further south.

'You were more than 500 miles apart. How did you keep in touch with each other effectively?'

'I would travel to Bingara annually to discuss progress, to reconcile accounts and to renew our friendship. We kept records of our accounts by writing to each other and I sent people up from Albemarle to help out, a couple of them were my nephews newly arrived from Ireland, we met regularly at Wilcannia and Fitzherbert used to come downriver to Albemarle on a steamer occasionally. It did not differ from anything I'd been doing for the last 10 years on the Darling, Blue. In many ways it was easier – there were regular mail services, telegraph services were extending to the frontier country and the Darling River had become an important highway as the paddle steamers connected with the ports and railheads of Echuca in Victoria and Morgan in South Australia. A bank had opened in Wilcannia and we kept an account there as well as the accounts we held in Melbourne.

Joe Phelps's work as a parliamentarian in Sydney was starting to reap benefits for people in the grazing lands. He helped to extend government services.

But aside from that, the most important thing about Fitzherbert and me is that we trusted each other. We were more like brothers than partners.'

'How did you decide to dissolve the partnership, Holas?'

'Before we started Bingara together, when I first met Fitzherbert in Melbourne, he was seeking help and finance with a station he wanted to own and manage with a partner. And I was looking at something beyond my association with the Phelps brothers. We were both single. I could remain as a silent partner and keep my post at Albemarle before I left the West Darling and moved to Queensland. But a few years after we started the partnership I married, and my views changed. Madam enjoyed the home she had and she protested at the prospect of removing to a district in Queensland even more remote than Albemarle – and she was there only for a few months – she much preferred the urbanity of northern Tasmania. So I became reluctant to move to Bingara and decided to keep my lot in with the Phelps brothers who had been very good to me. When I told Fitzherbert, he was pleased. That was wonderful. I expected his displeasure but he wanted sole proprietary.

It had been a grand affair for us both – profitable markets and good seasons. We had fat cattle on the road to Victoria most of the time. Fitzherbert had built a comfortable homestead, he had accumulated cash reserves and lines of credit and he was able to make me a generous offer for my share. He planned to marry and he wanted to establish a family business with his children. We formalised the dissolution of the partnership legally in 1881 and made formal announcements in the Queensland newspapers.

DISSOLUTION OF PARTNERSHIP. THE PARTNERSHIP hitherto existing between Nicholas Sadleir and Fitzherbert Brooke, trading under the style and title of SADLEIR & BROOKE, Graziers, of Bingara, Warrego District, Queensland,

is This Day DISSOLVED. All Accounts due to and owing by the late firm will be received and paid by FITZHERBERT BROOKE who will in future carry on the business in his own name. Melbourne, 1st April, 1881. NICHOLAS SADLEIR Witness to the signature of N. Sadleir – J. J. Phelps. FITZHERBERT BROOKE Witness to the signature of F. Brooke – Edward Trenchard.'

He paused. 'The sad thing was that Fitzherbert never got a wife and family to go with the station he worked so hard to develop.'

I read this aloud from an old newspaper clipping from 1889:

EDWARD TRENCHARD & CO. have received instructions from James Swift, executor under the will of the late Mr. Fitzherbert Brooke to offer for absolute and unreserved sale, at Scott's Hotel, Melbourne, on THURSDAY, the 26th September next, at three p.m., The Bingara Station, Comprising the Bingara and Bundilla runs together with 9600 Cattle, 127 Horses, Plant, Furniture, and Stores. The Bingara Run consists of 310 Square Miles. Bundilla square miles 234. Total 544 square miles. Bingara is completely fenced with 3, 4, and 6 wire fence, substantial and well erected. The other improvements are four paddocks with stock and drafting yards, comfortable Homestead, kitchen, store and out buildings, and kitchen garden. The water supply is abundant and permanent. Bingara has a frontage of thirteen miles to the Paroo and a double frontage of ten miles to the Yowah Creek, and is watered by the Menyabrici, Bingara, Dowalla, and other creeks. There are also large reserves in dams and waterholes, but what gives special and enhanced value to the run are six permanent springs which proved invaluable during the recent droughts. Bingara consists of Mulga country interspersed with salt and cotton bush plains and well grassed flats.

Bundilla.—The improvements here are huts, paddocks, drafting and stock yards, and southern boundary is substantially fenced for ten miles with three wires. The run is well watered by the Bundilla and Werie Ela Creeks, with large dams which throw the water back for a long distance. Bundilla consists of open salt and cotton bush, and Mitchell, blue, and barley grass plains, alternating with light belts of mulga.

The cattle are quiet and well bred, and compare favourably with the other herds of the district. The herd was originally formed from the best of Southern breeders, notably Messrs. Robertson Bros., of Colac, from whom Messrs. Sadleir and Brooke, the original owners, acquired their stud.

The auctioneers can confidently recommend the combined property to investors, and, as the station is now well stocked, the returns will be certain and constant. For further particulars apply to the Agents, 114 and 116 (late 66) Queen Street, Melbourne.

'Yes. It was a good place. But that was the end of things for Fitzherbert.'

'Why was the station auctioned in Melbourne, more than 1000 miles away?'

'It was the financial capital of the Antipodes, Blue. That was where all the money rested.

7

Anna Georgina and Quamby

The great-grandmother talks of her childhood in Geelong, her removal to her father's sister in England and her education there. She remains in England for 10 years. She regrets her return to Victoria at the behest of her father, but eventually settles to tutor children at Geelong. Great Grandmother Anna meets Nicholas in Melbourne. He is enchanted and entertains her and her father. They attend church. They correspond, marry in Geelong and move to a new house at Albemarle. Anna hates the loneliness. She travels with Nicholas to Melbourne for the birth of their daughter. Anna returns with her baby in unhappiness to Albemarle's isolation. Nicholas recommends the Phelps brothers buy Quamby Estate in Tasmania. They do. They ask Nicholas to move to Tasmania to manage it and to continue to manage Albemarle. Anna delightedly moves to luxury, civilisation, servants and conversation.

I was in Bristol, England looking at the SS Great Britain preserved in a dry dock and wondered if my great-grandmother could remember travelling on her. She and her future brother-in-law, John Sadleir, had used the ship to travel to Australia. I wondered how many other relatives had come to Australia on this luxury vessel. I asked her. 'Anna Georgina, can you hear me? I know you travelled on this ship in 1869 to re-join your father at Geelong in Victoria. What do you remember of the voyage?'

'And who are you? I guess from your accent you are Australian. Are you a relative of mine?'

'I am your daughter Georgie's grandson, Great-Grandmother. She had two sons one of whom was my father, Robert Raimond, and then she died. I'm sorry I never met her, I can tell you so little about her, but I did meet others of your children, Mary, Angela and Kathleen at Ballysinode, the house you had at the Grange in Adelaide.'

'Yes. That was our third Adelaide household. How well do you remember it?'

'It was in a fine position facing west to the horizon of the waters of the Gulf of St Vincent with a broad sandy beach in front and there was a jetty. It was several hundred yards long.'

'And the house?'

'It was a terrace house – built in a block of eight with two upper floors'. I went to the ground floor only where Great-Aunts Kathleen, Mary and Angela lived. They leased the upper floors to tenants.'

And we talked of her children, and the house and garden but she never told me of her voyage on the Great Britain. Perhaps it was unpleasant. She talked at length of her time in England.

'My mother died soon after I was born in Launceston in Tasmania. My father worked there as an assistant harbour master but I remember nothing of Launceston (I remember a great deal of it later in life). Pater kept a household in Geelong, he worked as a coastal surveyor for the government of the colony of Victoria, and those are my first memories. It was a bright, airy house looking out to sea and I had a loving nanny called Sheila Harrison. She had small children of her own: Margaret, who was about two years older than I, Herbert, who was the same and Robert, who was a baby, and she and her husband lived in the house with us. We had a cook as well, but Sheila was housekeeper as well as being my nanny. She was lovely.

When I was seven Pater sent me to Plymouth in England to stay at his sister's household and to go to school there. It was a terrible wrench. Perhaps you can imagine it – er, I don't know what to call you.'

Anna Georgina Sturgess, Geelong, Victoria, aged seven years

'Robert.'

'Good, one of my son's names. Perhaps you can imagine it, Robert. I thought my father was heartless and cruel and I cried for days when he told me of his decision and I begged to be allowed to remain in Geelong. Sheila tried to comfort me. At first, I flew into terrible rages, but little by little, she helped me to prepare myself. "Your father will not change his mind, Anna. You must try to think about how life will be with your aunt and your cousins in England. You already love books and reading. Your father has given me this letter from your aunt to read to you. She seems kind. Let us see what she is like." – and Sheila read Elizabeth's letter to my father.'

11 Citadel Road
Plymouth
England
4th of August 1859

My dear brother William,

Your letter received on 28 July gave me great pleasure. I am delighted with the descriptions of your daughter Anna and will do all I can to further her education and Christian care. I am to be married to William Cuddeford in late November and he has already warmly agreed to receive and welcome Anna to our household. William is a naval man but he does not go to sea; rather he works as a clerk for those who do and thus is not subject to the perils and adventures that beset most male members of the Sturgess family. I am pleased to say, we will have a calm and settled household, willingly and ideally suited to serve your daughter for the considerations you suggest.

I have found a school I believe will suit Anna. St Andrew's parish is establishing a class for girls and infants under the direction of Miss Hannaford, an instructress with strong Christian beliefs and an impeccable reputation for kindly teaching of arithmetic, grammar, spelling and needlework. The children of several respectable families are already enrolled there and so I expect Anna will be in good company. The fees are moderate and can be easily accommodated within the allowances you suggest.

I note your arrangements for Mr and Mrs William Sargent to accompany and care for Anna on her journey from the Port of Melbourne to Great Britain and your assurance of their continued affection and devotion to your daughter. The journey will be a wrench for Anna and it is kind that you have arranged for sympathetic and experienced guardians during the voyage. You tell me that you have not yet chosen a ship for the passage and can therefore not tell me precisely when I may expect to receive Anna and Mr and Mrs Sargent except to say that it will be sometime in January. We are, I must remind you, William, members of a seafaring family; there is no need to take an apologetic tone for the uncertainty of the time of debarkation. All that I pray is that you choose a good ship to *transport the entourage and pass an invitation to Mr and Mrs Sargent to stay for several nights in our household in Plymouth to assist in the welcome of your lovely Anna.*

I expect the trio will have to travel to me via coach from another port. There is little commercial shipping in Plymouth. I await their knock on my door. William and I will still be living at this address. Brothers Charles and William have been sharing a household with me, but they will be leaving soon for postings in Greenwich leaving a blissful and more suitably feminine household to share with Anna.

Your loving sister
Elizabeth."

I asked Sheila who were Mr and Mrs Sargent.

"Oh Anna, you call them Aunt Grace and Uncle William. They were here to dinner only two days ago."

I felt even more betrayed. To think that Aunty Grace, with the soft warm bosom and the sweet smell of cloves and cornflour, could have conspired with my father against me months ago! I was not surprised at Uncle William's betrayal. He and my father were like peas in a pod. They both smoked stinking pipes and drank brandy and got red-faced. One could expect betrayal from them! But Aunty Grace! She with the smiling eyes and the lovely storybooks!

I was still white-faced, fuming and silently contemplating revenge on "my best aunt in the world" when Sheila picked me up, put me on her knee and asked, "Well what you think of your Aunt Elizabeth?"

"Who is Aunt Elizabeth?"

"The writer of this letter."

"How can I tell what she is like from a silly old letter?"

"Well try to imagine. Do you think she is old or young, or fat or thin?"

"Old."

"How old?"

"Older than me. Old enough to get married."

"Yes, that is good. Fat or thin?"

"Show me the letter, Sheila." 'And she did.'

"Fat!"

"What, from the letter, made you think that, Anna?"

"The big fat droopy loops on her efs and gees."

"Ah, a clairvoyant as well!"

'And so it was that Sheila persuaded me to lose my resentment and look forward to the voyage – and she showed me some tricks that I've never forgotten about how to manipulate children. It was very useful in my large family later on.'

'And was Elizabeth fat or thin, Great-Grandmother?'

'Neither, she was well proportioned. When I first saw her, I thought she was astoundingly beautiful, dressed in fine clothes and surrounded by beautiful furniture, but it did not make me like her straight away. I still had my pride to consider – father, my betrayer, could not possibly have chosen a worthy sister to look after me. But I have to concede he did, and overall, she was lovely. She was kind but firm and the best thing about her was that she was very well-educated; she always pushed new and interesting books in my direction and she chose good tutors and good clothes for me. She was the closest thing I had to real mother, and when she started to have children, they became my brothers and sisters.

When my father called me home to Geelong 10 years later, the wrench of leaving my English family was as great as it had been when I left my father's household to come to Plymouth.'

'What were your first impressions of Plymouth and your Aunt Elizabeth's household?'

'Everything whirled Robert. The household was the most comfortable and familiar. There were glimpses of the sea and ships, and that seemed like Geelong, and so did the furniture and the routines, but the people of the street disconcerted me, they seemed furtive and rough and I could not understand the speech of many of the girls at school, and they thought I spoke strangely, and some of the more common girls teased me about my speech. I had lots of tears in and around the school in those first few weeks but Miss Hannaford noticed, and spoke to the girls about the "need to reach out in Christian charity to our cousins from over the seas". She was kind to me and she gave me special attention to make sure that I excelled above the others, many of whom came from ordinary working families and did not aspire to study of language and literature. I think Aunt Elizabeth had made it clear to Miss Hannaford that I was to undertake further learning in music, art and literature and somehow I felt like a special project to her. She was a wonderful teacher and protector; sadly, I was with her only for three years. Uncle William transferred to the Admiralty office and naturally, Aunt Elizabeth and I went with him.

I think my father paid his sister Elizabeth a generous allowance for me, because after we left Plymouth I did not attend school but had the services of governesses and tutors.'

'Did you not miss the company of other children?'

'No. It was entirely agreeable. My aunt arranged for me to join tutorials with other children at neighbouring households, and so there was a small group of children of my age who moved from household to household taking special instruction in Music, Latin, French, Literature and Classical History from those tutors who enjoyed and taught those subjects best. From time to time Aunt Elizabeth would persuade me to live in households where I would get special instruction from learned members. I was never far from her and she would look in to see to my welfare and keep me company several times weekly. She often brought some of her children. I loved to see them. They were like little brothers and sisters.'

'Well they were cousins.'

'Yes and the worst thing was that when I left Greenwich to return to Geelong I feared I would never see them again. And that was true except for one of them; Alice came out to Melbourne as a violinist with the visiting Royal Ballet Company and I was able to go to the performance by taking a steamer across Bass Strait from Launceston. I brought several of my children with their nannies and Alice and I were able to spend part of a day together at Royal Park. It was wonderful. I took two of the older children to the ballet and to see Alice playing in the orchestra pit and they both went to sleep in the box I had specially booked to help them to see my cousin performing. I learned after some years that children seldom live up to expectations.

Uncle Francis was in the Royal Navy too for a period and he worked with the Admiralty at Greenwich. He was a lovely jolly man and, for some years, he lived with us. He and Uncle William used to go off to work together each morning and they spent hours in the evenings having us in stitches with stories about some of the inept admirals they served. Uncle Francis came to Brisbane as a headmaster and many of my cousins were born there. I saw him rarely but we did keep in touch from time to time and Pater moved to live with him in Brisbane when he was very old. Pater died there, in Uncle Francis' house.'

'How did you decide to leave England, Great-Grandmother?'

'It was decided for me, Robert. I was having a fine time and so I resented a separation again. I was 16, and a lovely young midshipman was courting me. We were starting to fall in love. Aunt Elizabeth noticed. I think she was pleased and she sent a letter to my father breaking the good news. That was a terrible mistake. Pater ordered me home immediately. He sent me an obtuse and deceitful letter. I still remember it word for word.'

Geelong
5th of July 1869
My dear Anna,

I have disturbing news. The government of the colony of Victoria has reduced me in pay to half my normal salary and I can no longer support the expense of your education in England. As a consequence, you must return immediately.

I have written to your Aunt Elizabeth ordaining this and she and her husband William will find a suitable vessel for your homecoming.

I regret the nature and suddenness of this announcement, but I have little choice.

Your loving father,
William Sturgess.

'Why you think the letter was deceitful and obtuse?'

'Because when I returned to Geelong the household was thriving. Pater was not on half pay at all. He hosted many kinds of low fellows who spent most of their time in coarse conversation, smoking and drinking brandy on the front veranda and looking out to sea. There was lots of money for that. His actions in calling me home were blunt, stupid and cruelly deceitful. He knew I was unhappy, and my presence as a grim-faced, maturing, young woman rather than a simpering pupil seemed to unnerve him. He seemed to start at my sudden movements – say when I dropped a book or moved past him quickly on the veranda when he was entertaining his friends. I wondered if he was in poor health, but in the end, I had to have it out with him.

"Pater, please tell me the real reason for recalling me from England. You do not appear to be in straitened financial circumstances and yet you told me you were. Is there something else?"

"I simply wanted you home, Anna."

"Are you sure it had nothing to do with my connections in Greenwich?"

"I'm sure I do not know what you're talking about, Anna."

'And there it was, Robert. Deceit. But one did not call one's father a liar. Sheila had gone and there was nobody I could share my anger and grief with. It was a terrible time for me and all I could do was to write letters to him at Greenwich. However I never posted them. I feared that his life would be more unbearable if he heard from me in such unhappiness.

'What was his name?'

'I shan't mention it, Robert. There is no lingering love but it still makes me sad to say his name.'

She paused. She was silent for about a minute and, when she recommenced, her voice had changed. 'We started this conversation with a question from you about how I enjoyed my voyage out on the SS Great Britain. As perhaps you can guess, I was almost unaware of the ship. John Sadleir knew more about it – I can remember discussing it with him years afterwards. He was on its first voyage to Australia and I travelled on it about 17 years later. My main memory was my rage and disappointment at not being met at Port Melbourne. Pater had a good excuse, and he sent a messenger to collect me and take me to his offices in Melbourne where he was attending an important meeting, but it still grated. We travelled home on the new railway and we hardly spoke until journey's end in Geelong.

She paused again. 'But as with all things, I adapted. Within a month, I had found employment as a tutor and governess. I lived at Pater's house and travelled from household to household entertaining and instructing children in music, French and poetry. Later, most of the children came to me and Pater agreed to my setting up a special schoolroom beside a veranda looking out to the bay. I preoccupied myself with the work, I was independent and did not have to seek money for clothing and books from Pater (I would have hated that). I persuaded Pater to attend church again, he and I went to church every Sunday, and there were enough young people of a suitable class at Geelong for diverting friendships – but that was all. I avoided any serious engagements with anyone of my own age for several years.'

Holas told me how he met Anna Georgina.

'The 1860 and 70s were wonderful years in the West Darling. We had good to average seasons and we were learning about the country. We fenced it, sank wells and dams, built shearing sheds along the Darling, and developed huts and yards away from the river so that we could use more and more of the country available to us for grazing. John and Joe left the management to me. I moved regularly between Albemarle and Tarcoola and I was able to keep a good team on both places as well as helping Fitzherbert Brooke start a new station on the Paroo River. More people were coming to the West Darling from the goldfields, the paddle steamers were opening things up, Wentworth and Menindee developed as towns and ports, a good surgeon came to Wentworth, banks opened, and the riverboats made us feel part of the civilised British Empire with the ease of carting wool away and bringing us stores and luxuries. John, Joe, Nicholas Chadwick and I were justices of the peace and so we helped the police with their civilising influences. Regular mail and coach services ran from Wentworth and extended up to Wilcannia and Bourke.

Towards the end of the period, I think we shore more than 200,000 sheep with more than 500 people. We had more than 2000 cattle and up to 1000 head of horses.

Sadly, Nicholas Chadwick died in Melbourne in the midst of all this prosperity. He was young – only about 40 years old and it was a shock to us all. His executors sold Billilla, but Cannally remained with his sisters and nieces and nephews and it was managed by executors of his estate. We bought many of his stud sheep to start ram breeding at Albemarle.

John and Joe lived in the district for only part of the time. John lived at Tarcoola when he was in this country but he spent several years in Ireland managing his family affairs there and Joe had been elected a member of the Legislative Council for New South Wales. He spent a lot of time in Sydney and he travelled to America and Ireland and as well.

I needed to travel too. I went to Melbourne regularly to speak with our shipping agents and livestock agents and to look at markets – we shipped wool regularly for sale in London and most of our fat sheep walked to railheads for sale to Melbourne butchers. We had no telegraph then (it was a boon when it came to Menindee later) and for commercial transactions we relied on letters and face-to-face contact with people we trusted.

I always saw Richard on these trips to Melbourne (I usually took a stagecoach from Menindee to Wentworth, a steamer up river to Echuca, and the train to Melbourne). He and his wife Eliza had soirées often and they invited me whenever I was in Melbourne. But this was a special occasion. We were welcoming another brother from Ireland – Marshal and his wife Alicia were newly settled in a legal practice at Mansfield in Victoria and they had travelled to Melbourne seeking Richard's help with the imminent birth of their ninth child.

That was where I met your great-grandmother, Blue. Richard was keen on maintaining ties with old Ireland and somehow he had made the connection between the Hunt family of Tipperary and the marriage of one of their members to William Sturgess. He had been the master of HMS Pluto on the River Shannon before he came to the colonies with his new wife, born Georgina Margaret Hunt. I think Richard had been friendly with some members of the Hunt family before we left Tipperary and so he was keen to revive an antipodean family connection – perhaps it was a special kind of welcome for Marshal and Alicia – or it could be he just liked making new friends. So, for whatever reason, Anna Georgina Sturgess and her father William Sturgess were there.

Anna as a young woman

For me it was love at first sight. I was overwhelmed, constantly aware of her presence in the room (and she seemed always to be at a distance) and hardly able to sustain conversation with the bankers, surgeons and merchants and their wives who politely enquired about the seasons on the West Darling, the price of wool, the prosperity of paddle steamers and the price of shearing – all in an effort to make me feel comfortable and welcome. Finally, I was formally introduced. Eliza Sadleir did it. "Anna, may I present my brother-in-law Nicholas Sadleir from the West Darling district of New South Wales? Anna, Nicholas. Nicholas, Anna." Anna extended a hand. We talked, but I cannot remember what we talked about.'

'I can remember.' It was Great-Grandmother Anna Georgina, and she was giggling. 'We talked about the Crimean War because Nicholas had worked on the goldfields with several returned soldiers and there was a hutkeeper on Albemarle who had been a cavalry sergeant, and been wounded and had a wooden leg. Nicholas railed against anything Russian, and he conducted nearly all of the conversation looking over my left shoulder.'

'Really, my dear.' Nicholas could hear her. 'I thought we talked about Alfred Lord Tennyson's poetry – *The Charge of the Light Brigade* – and so on.'

I felt Nicholas had had the last word. Anna Georgina fell silent.

'Madam probably had a better memory of the conversation than I did. I felt like a nervous schoolboy rather than an accomplished pastoralist in his late-30s. And I was an honorary magistrate. I fear my lack of confidence betrayed me. I felt as if I was blushing. Eventually Richard came to save me by introducing me to a Melbourne banker and Anna Georgina moved to another group, but I remained aware of her position in the room for the rest of the evening. As the guests were leaving and lingering over their farewells I asked her if I might write to her inviting her to accompany me at the next race meeting at Flemington. "Please do," she said. "I think I should enjoy that. Richard and Eliza have my address at Geelong."

'And, Great-Grandmother, what happened after that?'

'A finely constructed letter arrived the following day by hand.'

Menzies Hotel
Melbourne
4th March 1873

Dear Miss Sturgess,
Please allow me to graciously acknowledge and thank you for your kindness in allowing me to address you in this way. I feel that we are kindred spirits, both with roots in the noble County of Tipperary – yours from the Hunts of Ballysinode and mine from the Sadleirs of Brookville and the Clarkes of Cashel. That much we know about each other, but I hope we have many more noble things to discover.

May I advance my prospects by describing my status as a gentleman in the colony of New South Wales and Queensland? I am employed by two brothers who are gentlemen from County Limerick to direct the management of their flocks of sheep, herds of cattle and troops of horses and to maintain the proper conduct of their stations adjoining the banks of the Darling River. There are more than 200,000 sheep, 2000 cattle and numerous horses. We employ about 500 souls who work as shepherds, hut keepers, well sinkers, fencers, drovers, bakers, butchers, cooks, housekeepers, builders, wheelwrights, shearers, wool sorters and overseers. We maintain two stations displaying most of the accoutrements of civilisation along the Darling River. Both are easily reached by paddle steamer connecting with the Port of Echuca and thence with the railway to Melbourne.

In Queensland, I share a partnership with a fine gentleman in a newly developing station in the Warrego district. We intend to run 10,000 cattle. Two of my nephews from Ireland will transfer there as my apprentices.

I shall be in Melbourne for the next three weeks. Will you and your father do me the honour of accompanying me to the race meeting at Flemington on Saturday next and dining with me at this hotel afterwards? I have arranged bedchambers for you both on the eve of the race meeting and at the conclusion of festivities.

I am conscious that visits from Geelong to the city of Melbourne are somewhat of an expedition for you. If you agree, I will travel by train to Geelong on Friday to collect you and your father at your residence, accompany you in a cab to the railway station and thence on the last train to Melbourne that evening. I hope we three may dine together that evening and choose a suitable amusement afterwards from the many that are on offer in the theatres of Melbourne. And following the races on Saturday we may repeat those pleasures? Perhaps on Sunday morning we may attend church together before you and Mr Sturgess return to Geelong on the train?

Miss Sturgess, please forgive my forwardness in proposing a prolonged period of social engagement. I have little time in Melbourne and I am anxious to attain your good offices so that I may benefit from the pleasure of your company before I return to the colony of New South Wales.

I await your reply.
Yours sincerely,
Nicholas Sadleir.

'Finely constructed, Great-Grandmother?'

'Well enough, but perhaps I should have said enthusiastic, Robert. And I was impressed by his offer to accompany me and my father to church. Many of the colonial men of Victoria neglected churchgoing.'

'So you went to the races?'

'Oh yes. And we did everything besides. He was generous. We had fine dinners and Pater left us free to go to the theatre on Friday night and on Saturday following the race meeting. He remained at table with his brandy on both evenings and it was pleasant for me to get away with a lively, alert, strong and purposeful man rather than put up with an often cheerless and depressing father. I had a fine new dress for the races. Nicholas was quite good-looking too and well turned out, and although he was much older than I, we shared an interest in contemporary literature (he kept a reasonable library at Albemarle he told me). We were even able to chatter in French. I think I was able to divert his attention from the horses for most of the time, and knowing Nicholas as I did later, that was a major accomplishment.

'So the social engagements were successful.'

'Most satisfactory, Robert. I don't think I fell in love with him in the way that I had with my midshipman in Greenwich, but he just seemed so good at everything, and so self-confident. Do you know that when he went back to Albemarle he immediately arranged with the owners to have a new house built at the homestead?'

'How did your courtship progress?'

'We exchanged letters. They became more frank and intimate and he came to Geelong late in 1873 to ask my father directly for my hand (he had already asked me, and I had agreed, but there was a formality in those days – one's father had to be acknowledged). Nicholas had a way with people. He disarmed my father who gave us his blessing instantly and put his arms around us both. I thought about it for years afterwards. It was the greatest display of joy I can remember from Pater. Nicholas proposed in lighter moments that it might have been Pater's expression of relief in getting rid of me. But I believe he was genuinely happy for us both.

It was a lovely wedding at Geelong on 25 February 1874.

Nicholas' three brothers were there with some of his nephews and nieces, (John's Australian-born, and Marshal's newly arrived from Ireland, as well as an Australian-born infant) and one of John's daughters, Melisina, was my bridesmaid wearing a new dress from Nicholas. Marshal, I remember, made a gracious and eloquent lawyer's speech that seemed to fall so prettily from the lips of the well-educated Irish. I felt much at home with this lovely family – a surgeon, a police inspector, a barrister and a pastoralist with a scattering of their families.

Nicholas regretted that his twin, Helena, was absent. Richard and John had met her when she came to Victoria about three years before seeking positions as a tutor or governess for children of gentleman. She remained in Victoria for only about six months and transferred to Western Australia for a position on a cattle station in the Gascoyne region as a governess. Nicholas wrote to her there several times and the letters were returned. John had tried his good offices as a Victorian inspector for cooperative efforts from the Western Australian police force but they had not been able to discover Helena.'

'A honeymoon?'

'We had it at Bendigo on the way to Albemarle. We spent the night at a Menzies Hotel in Melbourne and we caught the train to Bendigo and stayed there for three nights before we progressed to Echuca to catch a steamer going down river. Nicholas delighted in presenting me to a woman who ran a large emporium there.

"Mrs Edyvean," he said. "At long last I have pleasure in telling you I have acceded to your request. May I present Mrs Sadleir?"

She was lovely to me. "I'm delighted he has found you at last, my dear. And what may I call you?"

"Anna," Nicholas said.

"Nicholas, I instructed you to find me a Colleen, and you've exceeded all expectations by finding an Englishwoman."

"Not English, Mrs Edyvean – colonial-born, just English-educated, a splendid mix, and certainly superior to some of the Cornish folk we find in Ballarat and Bendigo."

And Mrs Edyvean threw back her head, roared with laughter, hugged Nicholas and me and led us into her parlour to be served afternoon tea. "This young man is not much improved from the time I met him, Anna. I expect you to report ameliorations to me on your next visit."

It was a hot February but we had the luxury of spending the first days of our marriage on the waters of the Murray River chugging downstream from Echuca and skimming past the beautiful river gums Nicholas had told me about in his letters. Each looked different. Nicholas told me the native Barkindji knew each individually at Albemarle.

Then we arrived in Wentworth in searing heat and had to wait five days in a hotel for a steamer going up the Darling to Albemarle (coaches were not running) and I hated it. Wentworth was a raw rough place. Black women were overly familiar with visiting shepherds as the men reeled in the streets with bottles of rum or brandy in their hands. It was hot and dusty. The language was foul and I did not know how to respond when Nicholas was not there (when he was with me people spoke properly). When we found a steamer going upriver, it had no suitable accommodation for me, but it was towing an empty barge and Nicholas persuaded the captain to rig a tent on it for us and furnish it with a bed and chairs, and so we were able to sleep and perform our ablutions in privacy, but it had something less than the cabins, bathrooms and lavatories we had enjoyed on the journey from Echuca.

Albemarle was new and strange. As we left the boat, everyone was craning necks to assess the new Mrs Sadleir. I felt like a freak in a circus. I was the only white woman for hundreds of miles. There were two black women who were my servants but I feared them and I wasn't able to tell them what I wanted because I didn't know what I wanted. The station cook operated in a separate building away from the homestead, the homestead was new, no one had lived there and so there was no routine for feeding the inhabitants. On the evening we arrived, after Nicholas had been seeing to some routine matters in the station store, I waited for him in tears in a partly furnished dining room with a table set with cutlery I found in a handsome sideboard, but with no food anywhere.

Nicholas comforted me. "I am sorry, my dear. I'd forgotten we needed to establish a routine. I will ask Nellie and Judy to bring us some dinner from the cookhouse. Tomorrow you and I will discuss matters with the cook and the servants, and with them we will start to establish you as mistress of Albemarle."

'I settled into a routine eventually and I discovered that the servants were kind and gentle and very fond of Nicholas. I think I showed my jealousy and when he became aware of it, Nicholas just laughed. "These people are like my children, Anna, and they behave like that, responding to kindness and praise. You are the only wife on Albemarle."

I enjoyed choosing wallpaper and planning the furnishing of the house. We had catalogues from furnishers in Melbourne to help. When we arrived, the builders were still finishing the house – fitting doors and windows and plastering some of the rooms. It was an adobe building – made of straw and mud compressed between boards to form the walls and it had a modern corrugated galvanised iron roof above twelve-foot ceilings. I think Nicholas got the idea for using this building method from a Californian who came to Albemarle from the goldfields and supervised the construction. The two-feet thick walls were made to keep the heat outside and retain coolness inside the building, but just as it worked to repel the external warmth in the beginning, during periods of prolonged heat, the huge walls helped the house to keep its heat, and so on cool nights in summer, the house was hotter inside than out. Nellie, Judy and I contrived a method of closing the house on hot days and drawing all the curtains, and reversing the routine to open all doors and windows to cooling breezes when night fell. This helped, but I never got used to the heat at Albemarle. And I was desperately lonely. I seldom saw other women with whom I shared any interests and I think the black women were frightened of me. Their children ran from me. The men of the station seemed not to know how to treat me, so mostly they ignored me and I was able to make friends only with the cook, the dairyman, the butcher, the gardener and the baker because I spoke to them about the food Nicholas and I needed, and even then, they did not always understand what I meant. Their ways were crude and simple and I was not sufficiently versed in household management to explain the level of elegance I wanted.

Sometimes I went for long walks along the river out of sight of the homestead and watched the absurdly large-billed pelicans fishing, and in the evenings I could see the wonderful red kangaroos coming in to drink. It diverted me for a while but it reminded me as well about how strangely different this country was.

It was pleasant when we had visitors to dine, even if the cooking was crude and the house unfinished, but that was rare. Joe Phelps stayed to see me and the new house for a day, he was courteous, polite and he seemed cold, the sheep classer stayed for a week, and Marshal's sons James and Stephen stayed for two days preparing to go to Bingara after an apprenticeship at the outstation at Victoria Lake (they were lovely – shy Irish boys, and about my age, but in awe of their Uncle Nicholas).

Nicholas and I usually dined alone and sometimes he was away for a night or so, so I was quite alone. I was lonely then but never frightened. Nicholas made me feel safe with the reliable way the station ran. People respected others and did their work cheerfully.

Soon I was pregnant. Nicholas was delighted and immediately made plans for my confinement at Richard's accouchement hospital in South Yarra. He wanted me to go to Melbourne months before the expected time of birth. I protested, I needed his love and support and he could not stay in Melbourne for the period I would be there, but secretly I was pleased. There would be white women I could talk to, not just about babies and pregnancy and childbirth, but anything at all that had nothing to do with sheep, horses or cattle.

Sam Curtis, Nicholas's old friend, the riverboat captain, called to Albemarle to collect us to take us to Echuca on a drizzly morning in August. It was cold but he sat me in a comfortable cane chair sheltered from the rain and radiated by the heat of the boiler with a tray of tea and freshly cooked johnnycakes at hand and he spread a blanket over my shoulders and another over my knees. "Welcome aboard Mrs Sadleir. I'm so sorry it has taken us so long to meet. I thought Nicholas would never marry. It is wonderful that he has finally met his match." And he continued to fuss over me. "I have given over both the visitors' cabins to you and Nicholas. I hope you find them pleasant, and please ask me for anything you may need."

Sam was the charming, rough, gentle rogue of the river traders but he was disarmingly plain and candid and he treated me like a daughter. He told me stories of his sister's children on the farm at Echuca, their progress in schooling and his ambitions for them. He asked my advice about the best means of tutelage for them and we talked about how he might improve his range of stock for sale for families along the river. I could see and feel the care he took to make me comfortable and welcome and Nicholas just smiled his thanks. Clearly, they were old, old friends (in bed on the first night out Nicholas told me the story of their meeting in Swan Hill and the way John Phelps encouraged Sam to extend his services up the Darling and Murrumbidgee rivers. Nicholas told me how Sam had come to the colony as a convict and Nicholas was proud that Sam had redeemed himself and prospered).

We called at Tarcoola to meet John Phelps. He was kind to me and warmer than his brother Joseph. "A belated welcome to the River Darling, Mrs Sadleir. I hope you come to love it in the way that your husband has. Is the new house to your liking?" I blushed and said I liked the house, and noting as I said it, that the house at Albemarle was far grander than the Tarcoola homestead, which comprised a collection of galvanised iron clad rooms with dirt floors connected by a covered walkway. I left Tarcoola feeling slightly embarrassed that I was not more grateful for the comparative luxury I enjoyed at Albemarle. I was secretly looking forward to the urbanity of Melbourne so I kept my misgivings about Albemarle and its life very much to myself. Not even Nicholas knew.

We got to Melbourne soon enough and I was struck by the cold wintery streets and people scurrying for shelter as we engaged a hansom cab to take us from Spencer Street Station to Richard and Eliza's home in South Yarra. It was the first time Nicholas and I had sat together in a buggy since we left Melbourne in late summer. It felt snug and comfortable inside the cab listening to the rain on the roof and the clip clop of the horse's hooves on a paved road. I was happy with myself and the growing baby within me. The sojourn at Albemarle seemed like a distant and slightly bizarre foreign holiday. This was home.

Eliza and Richard made me safe and welcome at South Yarra. Nicholas left soon afterwards on a steamer to Sydney and Brisbane for work on the Queensland station. I was sad to see him leave but I knew he would return to Melbourne before the baby's birth. I reunited with my father and spent several days with him at Geelong. We talked about my mother, and Pater's time with her in Ireland before they married in Dublin and came to Tasmania for work in the Port of Launceston. I had treasured a photograph of my mother Pater gave me to take with me to Plymouth and I still had it with me – well preserved, but Pater told me some other things I hardly knew. "Your mother was of a noble family, Anna. Her father, John Hunt was directly descended from the Earls of Oxford."

"Truly Pater? Why did you not tell me that before?"

"Is it important, Anna?"

"Well yes. I hope this baby will be the first of many. I want my children to have the advantages of life due to them. If they are nobly born it should be known and they should profit from it."

Pater responded by telling me that his letters to the Hunt family in Tipperary had drawn no responses for five years but that he would continue to try with branches of the family in Limerick to help me investigate my family history.

I told Pater about life at Albemarle. He expressed no wish to visit me there. There remained a considerable distance between us, but we were civil enough, and Pater invited my female friends to join me and I delighted in their company. I had been too long deprived of idle chatter. Before I married, I had a surfeit of it; I longed for serious discussions of poetry and philosophy from time to time, but now, back in the bosom of civilisation, I could not get enough of inconsequential gossip. I think Pater despaired at the hum of voices surrounding his daughter in his reception rooms at Geelong, but he put up with it for me, he continued to be a gracious host, and I was grateful. Even some of the children I used to instruct came to visit me.

Nicholas returned from Queensland via Albemarle 11 days before Eliza Georgina arrived. She was a beautiful baby, Richard attended me at the birth at his hospital in South Yarra, it was rapid natural birth with no complications and within six hours of my labour commencing, Nicholas was holding his new daughter. He had the look of a good father about him. We were both overjoyed.

Nicholas satisfied himself that I was well and that Eliza Georgina (we immediately called her Elsie) was thriving before he returned to Albemarle. I remained and spent Christmas with my brother-in-law and sister-in-law who were gracious in their hospitality and in their tutelage in the care of a new baby. They encouraged me to receive my father and he was delighted with his granddaughter. It was a lovely happy time but it had to end and I returned with Nicholas to Albemarle at the beginning of March. Elsie got a warm welcome, the housemaids delighted in her, I was happy to be with Nicholas again but I still missed the warmth and spontaneity of white female company.

I think Nicholas noticed my periods of longing and silence at Albemarle when I was not busy with Elsie. He left in a hurry without me for Melbourne, not promising a prompt return, but suggesting he may return with a pleasant surprise. I felt even more forlorn but he left in haste anyway, riding with a horse tailer and spare horses to get to Wentworth as quickly as possible. He told me that if a steamer was not available for Echuca to connect with the railway he would take a coach directly to Melbourne. His haste and mystery intrigued me and he avoided my questions with his hurry after his return from the telegraph station in Menindee. He left at dusk and hoped to be in Wentworth by the following evening.'

Nicholas continued. 'We had sold Tarcoola and were looking for other opportunities. Our agents in Melbourne had wired me that the Quamby estate was for sale. I had seen the estate years before when I visited cousins in Tasmania. I knew it to be wonderful country, of about 12,000 acres and there was no specified asking price. I was in Menindee, attending to other wires when the message arrived. I was able to send a wire to Joe at the Legislative Assembly in Sydney and to John at Albury where he was visiting his brother at the vineyards at Valverde. I wired them both: *Quamby Estate, Westbury, Tasmania for sale. Know property. Urge inspection. Nicholas.*

Joe responded the next day. *Inspect. Report Menzies Hotel 10 days.*

So I sped to Melbourne in three days, endured another bout of seasickness on the steamer voyage to Launceston and I interviewed the selling agent who met me at the Launceston docks as we drove out in the buggy to the estate on the morning of my arrival. It was 16 miles and we arrived in two hours. He was Mr Bell. He told me that the widow of Sir Richard Dry had the Quamby estate, but there was a large mortgage on the property (he didn't tell me how much but I knew it was £37,000) and Lady Dry wanted to remove to England to live there are permanently.

I read documents describing the property and a newspaper advertisement:

THE ESTATE OF QUAMBY.

J. BELL AND WESTBROOK have received instructions from Lady Dry to sell by auction, at their mart in Charles Street, Launceston, on MONDAY, the 19th day of April next, at 12 o'clock noon (unless the same shall be previously disposed of by private contract), THE ESTATE OF QUAMBY, containing 11,800 acres of land or thereabouts, and forming, beyond all question, the finest property in Tasmania.

This unrivalled estate, which is situate on the Westbury Road, and distant about 15 miles from Launceston, comprises a large portion of the richest agricultural and grazing land in the western district, having a frontage of upwards of 20 miles on the Meander River, and being also intersected by Quamby Brook. It is admirably adapted for the breeding of stud flocks, being copiously watered and abounding in rich natural and artificial grasses.

Quamby House (the principal dwelling) is a large and commodious mansion, replete with every convenience, and the garden and grounds, which are of great extent, are stocked with the choicest trees, shrubs, and plants which could be procured. The Hagley Station of the Launceston and Western Railway is on the boundary of the estate, and distant only two miles from Quamby House.

It is believed that when the farms which are now under offer are rented, the rent roll will considerably exceed £4000 per annum, and that with the steady and rapid rise which is taking place in the value of property in Tasmania, will in a short time be susceptible of being largely increased.

Possession of Quamby House, garden grounds, and deer park can be given at the expiration of one month after the sale.

Further particulars and terms can be ascertained on application to
MESSRS. RITCHIE & PARKER,
Solicitors, Launceston.
TUESDAY, 10th of February

I studied a list of tenanted farms, the description of the farms and their improvements, the amount of annual rental for each farm and the names of the tenants

"Mr Bell," I said, "You describe the estate as comprising 30 or so tenanted farms with room to develop more. Why do you not offer the farms separately? Surely it would be more profitable affair?

"I agree it would, Mr Sadleir. But that is not Lady Dry's wish. Many of the tenants are anxious about losing their farms from the sale and I think they have prevailed on her to preserve things as they are, and to sell the estate in its entirety as it is currently managed. It is not wise for me to proffer this advice as a selling agent, but I fear that those provisions will limit the estate's attractiveness to a range of purchasers. Many will not want to clash with the conservatism Lady Dry has encouraged the people associated with Quamby to expect. In addition, the permanent, firm direction of the actions of more than 30 tenant farmers who have enjoyed a loose rein for some time is no small thing. The straightness of management has been in decline. I hope, for Lady Dry's sake, that there is something left for her from the proceeds."

"Are the farms all let?"

"No. Some come up for renewal each year as their leases expire and the length of the leases vary from a year to 14 years. The Dry family varied agreements to accommodate the needs of their tenants. That is why in the 11,800 acres for sale I cannot specify exactly the number of farms available for lease. Negotiations continue all the time."

I said. "Mr Bell, what would you expect the 10 best farms to fetch per acre if you were able to sell them individually?"

He stopped the buggy for a minute to think about his response. "Between nine and ten pounds an acre."

"And the five worst?"

"Seven pounds."

We arrived at the homestead. The buildings were in good repair and they were clean and tidy. That was a good sign. It showed those associated with the homestead and the buildings run by the estate had some pride in their work. We walked through the house. Lady Dry had vacated it to live in Hobart but it was clean and well-kept with fine furniture and an adequate household staff was in evidence. The baker, butcher and household cook were still doing their work to support all the staff of the estate. Mr Bell told me the domestic staff numbered more than ten. Some came from the families of tenant farmers and others lived in quarters at the homestead. In many ways it reminded me of some of the big houses in Ireland. It had a similar feeling to Brookville House, our old Tipperary home, but the style was different – Brookville was square and block-like, Quamby was elongated with lots of outbuildings appearing to lean against it; and it had a long veranda facing the garden with glazed french windows leading to the salons within. It had an upper storey with rows of bedrooms but the upper storey wasn't obvious from the outside. It reminded me of some of the pictures I had seen of grand houses in the southern colonies of North America. Mr Bell agreed. "Yes." he said. "The architect was American. But I expect you realise that the main house is the last building in this cluster. Many of the buildings surrounding it go back to the first years of the colony of Van Diemen's Land. We have details of the architect of the main house, but I do not have them with me."

"And the builder?"

"Nothing I'm afraid, although I may make further enquiries of Lady Dry. I expect you are making enquiries of the supervising builder. All the tradesmen would have been convicts. There may be lists of them but I doubt it."

We left the homestead with its huge surrounding garden and deer park and walked to the farm buildings passing a neat two-storey manager's house. Built of stone and roofed with slate, the steading had most of the amenities one would expect of a small village – certainly outshining Albemarle's in volume and style. There were large stables, a shearing shed, a huge slaughterhouse, granaries, a dairy to make butter and cheese, a piggery, fowl houses, smokehouses for bacon and hams, and airy storages and cellars storing anything from last season's apples to maturing cheeses. The buildings reminded me in their substance and style of those serving the demesne[1] lands surrounding the big houses of Ireland, and I mused that this was as close to that as I was likely to see in the new colonies of Australia. I felt at home. Nearly everything that grew at Quamby was British – rows of deciduous fruit trees, newly harvested, fallow land awaiting turnip and wheat and barley seed and new green meadows with clover and grasses nourishing fat British breeds of sheep and cattle. And I smiled at the deer gently grazing – not a kangaroo in sight.

We took the buggy to explore the perimeters of the estate. Frank Sams, the estate overseer, came with us. I liked the cut of him. He carried himself like a sportsman and dealt with us respectfully but not obsequiously. I judged that he answered my many questions truthfully and with a background of understanding and care for the information he gave me.

"What horses do you keep, Mr Sams, and for what purposes?"

"We have more than 50, Mr Sadleir, two working six-horse teams, 12 general carriage horses and hacks, seven general dray horses and a troop of up to 15 breeding mares of mixed sorts – Clydesdales for the teams, some blood horses for hacks and for sport and some heavy types for all-round breeding. We keep an entire Clydesdale to service our mares and his services are available at a fee to our tenants and neighbours."

"And how can the organisation of the estate's horses be improved, Mr Sams?"

"I think we need more spare working horses for the teams. We overwork certain horses at busy times of the year and we need more to replace them and rest them."

"And how would you like to see that achieved?"

"Immediately I think we needed least seven more draught animals to work in the teams, and I think we should buy those with sums we can realise on some of the buggy horses of which we have an oversupply. With that correction, I think we have sufficient heavy mares to breed our own draft replacements for as far ahead as I can see."

"And what of the sheep? How many do you run, and for what purpose?"

We talked for hours about sheep, cattle, and rotations of cropping as we moved around the estate, noting the tenant farms and their well-kept buildings, the wooden post and rail fencing and hedgerows. We skirted the Meander River and called at a tenant farm Sir William Dry[2] had assigned to St Mary's Church at Hagley. He had paid for the building of the church too.

Mr Patterson, the tenant of the Church farm, had a lease of 14 years. He, Mrs Patterson, and several of his grown sons gave us afternoon tea. They were champion ploughmen and the mantelpiece of the large farm kitchen displayed shining cups from local and intercolonial ploughing matches. Mr Patterson and his sons described their farming programme. They grew crops on about a third of the farm and the remaining two-thirds they left as fallow or meadow. They grew English and Cape barley for malting and stock feed, wheat for milling (with yields of up to 50 bushels per acre) and potatoes and turnips, but it was not an entirely easy conversation. Frank Sams looked at the floor uncomfortably during some of the exposition and Mr Bell avoided my gaze. I asked general questions only about their crops and fertilisers and the horses they used.

As we left, Mr Patterson made an assertion about the poor maintenance of buildings provided by the landlord in comparison with the annual rental he paid to the estate. I shook his hand warmly and thanked him for his hospitality. I made no reference to his statement about the buildings or his rental but I noted it. It seemed to me not a special thing for a tenant to make a case for a more favourable treaty to the agent of the potential owner. I thought no worse of Mr Patterson but Sams and Bell were anxious to raise the matter as we drove away.

"I fear Mr Patterson may have left a bad impression, Mr Sadleir." Bell said, and Sams nodded and looked at me anxiously.

"In what aspect?"

"He understated some of his yields and overstated others, and his assertion of unfair treatment from the estate of Lady Dry seemed ill-considered."

"Yes," I said. "But I would have been surprised if he did

otherwise. He is obviously a fine farmer and trader. If my principals become buyers, he will have advanced his case in treating with them about the conduct of his leased farm and his affairs. There is nothing underhand about him that I can see. And remember he is a good farmer and he gave me information I can verify this evening with available statistics I have already assembled from the colony of Tasmania's annual reports."

Mr Bell and I thanked Frank Sams warmly as we left him near his mother's homestead at Hagley and proceeded to the wharf at Launceston. My steamer departed at five in the afternoon and I planned to spend the night appraising statistics and information for John and Joe Phelps whom I expected to meet on the day after I arrived in Melbourne.

I passed most of the journey back to Launceston silently thinking. Bell seemed not to mind the lack of conversation. I think he knew what I was doing. At the end of the journey I had concluded that John and Joe could pay up to £70,000 for the estate and have a profitable affair, but I knew that I should verify that with great care before I advanced that opinion to them. I asked Mr Bell. "You plan an auction soon here in Launceston. What value do you place on the estate? And might you persuade Lady Dry to accept an offer before the auction?"

Bell waited for me to descend from the buggy. "You pose a reasonable question, Sadleir, but I prefer to conduct an auction to answer it. However, should you offer £70,000 to Lady Dry I believe that we as selling agents, and combined with her attorneys at law, could persuade her to close on that offer."

I nodded. We shook hands and I boarded the steamer. It left for the Port of Melbourne twenty minutes later.

It was a rough crossing and I did no work that night. I retched and vomited across Bass Strait but I was recovered and steady as we berthed at Melbourne in the morning. I took a cab to Menzies Hotel, breakfasted, bathed and began work on my estimates of the value of Quamby. When John and Joe arrived the following morning, I was prepared.

"Cousins," I said, "two days ago I spent more than five hours at Quamby estate looking at the land and buildings and extracting information from the estate's overseer about its conduct. I have questioned the selling agent and I interviewed one of the best farming tenants. Yesterday and last night, I studied official statistics of production from the region. I made several calculations about the financial conduct of the estate. I conclude it has not been well-managed, that tenant farmers by and large have exploited the owner's lack of attention to prudent conduct, and that gentle improvements, if they are made in good faith, but firmly, will provide profitable returns and may double the value of the holding in a period of 10 to 15 years.

Joseph said. "You seem slightly hesitant, Nicholas. Why do you say *gentle improvements* – and *made in good faith?*"

"The tenants have a prevailing idea that the land they lease comes to them with enduring rights ensuring their family's occupation on generous terms. They benefit from profitable tenure, and that is favourable for them and for a future purchaser, because they are good farmers and encouraged by their profitability, but I think the present owner, the widow Lady Dry, has encouraged them to think that she will endow them with ongoing goodwill and prosperity. There is some sign that it may become an enduring northern Tasmanian belief and become a political platform if it is not properly dealt with by a new owner."

"What do you think should happen, Nicholas?" John said.

"I think the new owners should immediately give notice to all tenants that current rents agreed to with the present owners will be respected, but that rents will be reviewed and revised immediately to assist the landlord and tenant to plan future leases and their commercial consequences, and that it be done in turn with each tenant."

"Do you think you could do that effectively, Nicholas?"

"Yes Joseph."

"Have you valued the property Nicholas?"

"Yes. Farms neighbouring Quamby have sold recently for more than £10 an acre, and I believe if they are properly managed by a hard-working family, each acre will return a profit of 20 shillings a year – thus a 10% yield to capital. I estimate the whole of Quamby, valued at an average of less than seven pounds an acre should easily give 10% per annum after all expenses. Thus, it has a theoretical value of more than £100,000."

"We should have to borrow money to pay that Nicholas. What do you think Lady Dry will accept?'

"It is for sale by auction next week, but the selling agent suggested to me that Lady Dry would accept an offer before the auction of £70,000."

"Does that mean that the auctioneer thinks it will fetch less than that?"

"I believe so."

I stood to move back from the table we were sitting at, unpacked a satchel of papers and passed them across the table to John and Joseph. "These are background papers gentlemen, and some outlines of the logic I used to reach the recommendations I've given you. I believe Quamby at £70,000 or less will be a very good affair but you need time to consider that and raise the capital if you need to. I have no available funds. I have invested heavily in an enterprise with Fitzherbert Brooke of Bingara on the Paroo as you know, but I am happy to manage the enterprise for you if you wish. I need to walk now to clear my head, and you need to consider your choices. Perhaps we may meet for lunch within a couple of hours to talk some more – at say one o'clock this afternoon in the dining room?"

They smiled and nodded. They were already studying the papers I'd given them.

The Brothers Phelps were enthusiastic at lunchtime. They had decided to take my papers and look for themselves and they were leaving the following morning on the steamer for Launceston. They asked me to wait for them in Melbourne. They would return at the earliest within three days, and at the latest, within five.

I wired Anna at Menindee (I had instructed that the station buggy to go in each day to Menindee to look for telegrams from me) *Delayed Melbourne another week. Will contact you when plans are known. Love, Nicholas.* I knew it told her little, but it would be some comfort to her in her isolation.

Two days later I got a wire from Launceston. It said: *Will you manage it for us beginning immediately? Depending on outcome of sale?*

I replied: *Yes*

A further two days later John and Joe sent me a wire saying they had purchased Quamby and they wanted me to reserve time for talks the following morning after their arrival on the steamer from Launceston.

John was direct. "We agreed with your observations Nicholas, and we purchased Quamby for £60,000 at the auction. We have arranged to continue employing all staff. We want you at Quamby to install lasting and sensible systems of direction and accounting with Anna and your new daughter as soon as you can reasonably remove from Albemarle. What do you say?

"Am I to relinquish my responsibilities to Albemarle?"

"No. We want you to supervise the conduct of Albemarle and Quamby. For the interim I will remove to Albemarle with a view to installing a sub-manager there for you to supervise as you visit periodically from Tasmania.

"And my interest in Bingara and support for Fitzherbert Brooke?"

"Those are your affairs Nicholas. However, we would encourage you to review your independent commitments in our favour. To that end, we offer you a salary of £1000 to work intensively on improvements at Quamby and to oversee the good conduct of Albemarle. We will manage our Victorian investments."

I paused thoughtfully and Joe said, "I know this is sudden, Nicholas, and that you are newly married, but I sense that you are personally disposed to life in Tasmania and that you may need only a few weeks to persuade Anna to remove there. I think it will not take long. I have taken the liberty of obtaining some photographs of the homestead with lists of the household furniture and equipment for you to take to her. I imagine to a young mother of Anna's refinement it will compare more than favourably with Albemarle." And he raised his eyebrows with a grin. He was not a politician for nothing.

I stood. "You know me, gentlemen, and thank you for thinking of Anna." I reached across the table and shook hands with them. I think I just have time to take the train for Echuca. I will send you both a wire about our plans within a week". And I ran to my bedchamber to collect my bag to find a cab to Spencer Street Station.

It took less than three minutes to have Anna favourably disposed. Anna was feeding Eliza when I walked into the homestead and she started when she saw me and then burst into tears. "Nicholas, I've missed you terribly. I'm not sure that I can endure life here."

I stroked her hair and took the baby from her. "Then perhaps we should move. Would you like to be closer to your birthplace?"

"Tasmania! I'm sure I have been there because Pater told me so but I can't remember it. I was a baby when we left. Is it as remote and lonely as here?"

"No, it is lovely. I have some photographs of your new house for you. It is less than twenty miles from Launceston."

"Mater's grave is in Launceston. I have never seen it."

I held our gurgling Elsie as Anna examined the photographs of the Quamby homestead.

"Nicholas you're teasing. This is a fairy tale."

I remained silent. She looked up and smiled.

"When can you be ready to leave, Anna?"

"For how long will we be there, Nicholas?"

"Perhaps forever, my love."

"But what will happen to the homestead here? The decoration and furnishing is unfinished."

"John will come to live here as soon as we leave."

She seemed flustered. "And when will that be?"

"That is my question to you."

Anna left the room with her hands held high and shaking them. Within a minute, she rushed back.

"We can go there whenever we want?"

"Any time after the 19th of May. Lady Dry requested a month's grace before she gave up possession."

"Lady Dry? Who is Lady Dry?" And she turned to me with outstretched arms. "The 19th of May! But that is less than three weeks away."

Frank Sams met us at the dock at Launceston on the morning of 25 May 1875 and we drove to Quamby to resettle. My two ladies were content.

[1] Land held by the owner of the estate for personal use, and not leased to tenants.

[2] Sir William Dry was the first native-born Premier of Tasmania. He died young, left no children and his widow inherited his estate. His father developed Quamby and its gracious homestead with 30,000 acres. He was an emancipated convict who was transported to Tasmania because of his association with the Young Irelander Movement.

8

The Quamby Years

It is hard to imagine Quamby not providing the best of Nicholas and Anna's life. Eleven children arrive. They entertain lavishly. Bishops and governors dine. Five hundred enthusiasts of the Northern Tasmania Coursing Association watch and compete on Quamby and friends and relatives visit. Nicholas and his brothers discuss the Kelly gang's imagined connection with Fenians dynamiting London as a means to home rule in Ireland. Their mother in Ireland fears it. Marshal writes to reassure her and the brothers resolve to find their sister Helena who is missing. Anna starts a licensed school for her children. Nicholas has a salary of £2000. He directs Melbourne agents to finalise the estate of John Phelps who dies in Florence. He visits Albemarle, Laverton in Victoria, Valverde at Albury and starts new cattle stations in Queensland. He talks of frontier murder and massacre near Cloncurry and long cattle drives to stock the new stations.

Great-Grandfather described their Tasmanian arrival:

'Sams drove us promptly to Quamby from the Launceston docks on the morning of our arrival. He loaded some of our light bags behind us and left the rest of our heavy luggage to Tomkins who was there with a dray and came to Quamby from the town later in the day. Sams greeted Anna proudly and well. "Welcome to Tasmania, Mrs Sadleir. I am Frank Sams, overseer of the Quamby estate. I shall be pleased to assist you in any way in your installation here and in your management of the household. May I hold your daughter as you get into the buggy? We will proceed to Quamby directly and should be there within two hours, but perhaps you seek refreshment before we set out?"

Anna looked to me and back to Sams. "Thank you, Mr Sams. Baby Eliza is comfortable, Mr Sadleir is well fed and I am in no need of further nourishment. Should we carry water for the journey, or are there inns along the way that will serve refreshments?" She passed the baby to Sams and stepped into the buggy. Sams lent over the baby sheltering her from the rain before he passed Elsie up to her mother in the shelter of the buggy.

"We have even better, Mrs Sadleir. Cook has given us a picnic hamper for the journey. There is cake, lemonade, water and milk and I may boil a billy for tea if you decide to have it."

Anna nursed Elsie closely as we drove from the dock and passed through some of the broad streets of the town of Launceston. It was a cool day and raining slightly but I could see she approved of Launceston.

"This looks a fine town, Nicholas," she said. "I think I will like it as well as Geelong."

Frank Sams directed the buggy along the main streets and pointed out the Post and Telegraph station, banks, mercers, an ironmonger, the courts, legal chambers and markets.'

Great-Grandmother continued: 'Mr Sams seemed a cheerful fellow. He called to people in the streets of Launceston as we drove through it and many of the people stopped to stare at us. It was almost as if we were expected and people knew who we were, or they imputed that we were the new occupants of Quamby because of our presence in the buggy with Mr Sams and that Mr Sams alerted them to that by calling to them as we passed by. Their stares were neither hostile nor welcoming – just curious, but even so, I felt slightly affronted, or perhaps I should say alerted to the need to behave cautiously. I'm not sure that Nicholas even noticed. He and Sams spent most of the journey discussing the livestock, pastures and crops we passed on the way.

Eliza slept for the whole journey, we felt no need to stop for refreshment, the horses were willing and proceeded ably, and we trotted directly to Quamby passing through the villages of Prospect, Hadspen, Carrick and Hagley on the way. The landscape gave a pleasant outlook. There were none of the long searing landscapes or the endless horizons of Albemarle. This was a neat gentle district. The villages evoked a feeling in me of country England. There were neat inns, flour mills, tradesmen's premises of most sorts surrounded with neat orchards and gardens.

Quamby House, near Westbury, Tasmania

I was abashed at the sight of Quamby. It seemed far grander than I expected and we passed through acres of garden along the driveway with deer grazing in the distance. As we drew up to the front door, the staff of the household filed out. I paled to see them all – I think there were 11 – and Eliza started to cry. I remember distinctly feeling that I was not at my regal best, but Mrs Flood, the housekeeper, left the congregation on the steps and rushed to my side of the buggy to reach up and take the bawling Eliza. She could not have been kinder. "Welcome, Mrs Sadleir. Let us go inside to see to this wee tot." And she showed me to an airy private chamber laid out with water and towels where I could feed Eliza and clean and change her. The household staff parted to make a pathway for us as we rushed through with the howling infant.

Not only had I been saved from the ordeal of formal introductions with the household staff, but I could luxuriate in the physical pleasure of feeding my baby in private comfort and with a kindly senior woman overseeing it.'

Nicholas continued: 'Frank Sams and I stepped down from the buggy and Percy Brown unloaded our bags and led the buggy and pair away towards the stables. I said, "Thank you, Mr Sams. As we can see, Mrs Sadleir is incommoded at present; I'm sure she will be ready to meet the household staff later, but in the meantime, could you please present them to me?" And so Fred Sams and I walked up the line and met each of the household staff one by one. Most of them were nervous and a few would not meet my eye when I shook hands with them, but I expected that. Why would they not be nervous? Some of them had served the Dry family for two generations, and from what I could hear, they had been generously treated. Now they had new masters. Things would certainly be different. I made a short speech.

"Ladies and gentlemen, thank you for taking the trouble to meet us so formally. As you know, I am Nicholas Sadleir and my wife, Mrs Anna Sadleir, will meet you later as she proceeds to discuss your duties with you and the housekeeper, and to present our daughter Eliza. Mrs Sadleir and I intend to make our home here. We look forward to your friendship and assistance in the hard work we have ahead to maintain the pride and traditions of the Quamby Estate for the new owners John and Joseph Phelps, with whom I have been associated in pastoralism in the West Darling district of New South Wales for 16 years. I hope you will help us to entertain them at Quamby in fine style in the coming months and help us continue the tradition of fine hospitality at Quamby for the people of northern Tasmania. Thank you again for your welcome. Please return to your duties."

'Mrs Flood, the housekeeper had taken charge of Anna, or perhaps it was the other way round, but our first luncheon at Quamby was a delicious success, baby Eliza was already being cared for by a nanny, and was sleeping, and Anna was preparing for a meeting with the household staff in the mid-afternoon. Sams and I went to the estate office and started to study the accounts of the estate.

Anna and I dined alone in style that evening in the commodious and elegant dining room.

Maggie, one of the housemaids, served us coffee in the salon, asked us if "that would be all for the evening?" and withdrew.

"Well, Madam, how do you find your new circumstances?"

"Mrs Sadleir finds her new circumstances most agreeable indeed, Mr Sadleir, and I would prefer for you to continue calling me Madam at all times. It seems properly fitting." And she smiled and chuckled. I realised then how difficult life at Albemarle had been for her. Beautiful women are strange delicate creatures, Blue'.

'I know.' I said.

'Madam continued her happiness. The household staff adopted her and Eliza. She charmed the gardeners, the coachman and ostlers and even got a smile out of the dairyman, Sean Ryan, a man of uncertain age and unreliable temperament, from whom most of the kitchen staff had not been able to get a civil word for years. Anna was a transformed marvel of tact and good humour, she became active in matters to do with St Margaret's church at Hagley, and in no time at all she was hosting meetings for women of the district concerning projects extending Christianity, learning and goodwill to all people of the district. She took frequent buggy trips to Launceston to shop and sometimes she used the railway from Hagley. She was a guest at households in northern Tasmania when I was occupied during the day, she helped several households to establish libraries of new works from England and we were frequent guests at dinners and levies together in Launceston when we stayed overnight in hotels. And her father was a frequent visitor.

It was a delight for me to see Anna so happy, but I had serious work. For the first six months Frank Sams and I met every morning after he had assigned tasks to men – shepherds, ploughmen, herders and general labourers who worked on the estate not associated with tenanted farms. I listened and asked questions and we discussed solutions to problems together. We agreed on plans to import improved entire horses, new British Breed rams and improved strains of cattle for beef and milk. We talked about cropping plans and agreed on the ratio of potatoes, wheat, oats, barley and turnips for sheep feed to plant in the following season, and we made plans to import increased quantities of guano fertiliser.

That comprised my mornings' work. Each afternoon I set aside time to visit each of the leased farms in turn. I had written to each of the tenants proposing a time to visit them and inviting discussion on all matters to do with their leases and the conditions of the farms they leased. Generally, the tenants and their families were wary. I expected that. I'd written to each tenant hoping to mollify their fears and I should have known better. It probably made some families more anxious, and not all of them could read. I wrote this to them all:

I represent the new owners of Quamby, Messrs John and Joseph Phelps of Albemarle, New South Wales, and I seek an audience with you to discuss all matters to do with the farm you lease from Quamby estate. In proposing this interview, I seek friendship and frank discussion on further investment to provide necessary improvements and a fair annual rent matching investment in those improvements to assist the proper husbandry of the land you lease.

Nicholas Sadleir
Manager

In no time I found that the ways of the West Darling served far better than the formality of written letters. I took cheese from the Quamby dairyman as a disarming gift, and when many tenants began a prepared speech about a fair annual rental, I interrupted. "Man, it is far too soon for that. I need to know you and I need you to tell me what this country will grow what plans you have for it. If we understand and agree on that together, we can talk about what you need to do that and how we may find the will to furnish it. You are important to Quamby. You and your family form part of its history and prosperity. Please tell me the story of your family."

And that is how we generally began. Sometimes we explored our mutual roots in old Ireland and we were able to exchange a few words of old Irish (although my Irish was very rusty, and I never learned it properly anyway because my mother discouraged its use in Brookville House – I think she thought it was a akin to swearing, so I spoke it only in the fields when I was playing with tenant boys). Depending on the tenant, I often left our first interview with nothing at all agreed, only an appointment to meet again, "after harvest" or "after shearing" or "after sowing".

Most of my first meetings were with tenants who had leases with years to run. There was no urgency to our talks. The rent was set. We talked about future improvements in the best systems of farming for that particular farm and they were generally enjoyable conversations. I think the word got about, Blue. By the time I got to tenants whose leases were due for renewal, they were ready to talk about all aspects of the farm and not just the rental money. I learned more about farming in northern Tasmanian in that first year than I did in the rest of the time I was at Quamby.

But the greatest fun was greyhound coursing. Many of the tenants kept greyhounds and had suggested that I make some of the untenanted land at Quamby available for the sport of coursing. We had a good population of hares. My brothers John and Marshal and I had seen dog handlers and slippers at coursing competitions in Tipperary. It seemed feasible. Many proponents of the idea of were good jolly fellows, so we formed a coterie to establish the Northern Tasmanian Coursing Club. This was a newspaper notice of the first meeting in 1878. We had courses before then, but this was the first formal occasion and later the club became the *Northern* Tasmanian Coursing Club (the Hobart people thought they had a prior claim to *Tasmanian*):

The Cornwall Chronicle – Wednesday 5 June 1878
TASMANIAN COURSING CLUB.
The first meeting of the Tasmanian Coursing Club is to take place over the Quamby paddocks to-morrow, and is likely to be a most successful meet. His Excellency the Governor will be present, and Mr Sadleir, manager of the Quamby estate, has made all arrangements for a most enjoyable gathering. We understand there will be a goodly assemblage of sportsmen, and that the fair sex will be well represented. Mr Lord, manager of the Launceston and Western Railway, has provided ample accommodation for visitors. A train will leave the Launceston station for Hagley at 7.00 a.m., and the usual train at 8 a.m.; and return trains will leave Hagley at 6.7 p.m. and 6.46 p.m.

A considerable number of visitors from Hobart Town are expected to be present at this meeting, and, should the weather prove at all favourable, there can be no doubt the first coursing match in Tasmania will be a most successful affair.

This started a tradition for Madam, the children, me, and most of the staff that went on for years. We hosted the whole thing and fed everyone in the gardens of the main house. It became a major event. In one year we had more than 600 guests.

We bred dogs at Quamby. John had good dogs at his country postings with the police in Victoria, he would send up puppies for me to collect on my trips to Melbourne and we used them to improve our lines. Because I was the patron, I thought it was improper for me to put in entries, so the staff at Quamby and some of the tenants did it in their names.

Madam and I received memorials and flowery letters of thanks in the newspapers. Here is an example with our response.

Launceston, 9th July, 1880.
N. SADLEIR, ESQ.,
Quamby.
Dear Sir,

A large number of gentlemen who take a great interest in coursing think the time has arrived when they would like to recognise in some tangible form your kindness in allowing the Northern Tasmanian Coursing Club the use of Quamby estate for holding their numerous meets, and also for the hospitality both Mrs Sadleir and yourself have at all times shown to the public generally. In accordance with their wish, therefore, they beg your acceptance of the accompanying memento of esteem, and sincerely trust that you will receive at the same time the heartfelt good wishes of the subscribers to this testimonial, not only for yourself, but also for Mrs Sadleir and family. In conclusion, they hope that you will all be spared to take part in many more such social gatherings as have hitherto been held at Quamby in connection with the Northern Tasmanian Coursing Club.

We are, dear Sir, yours truly (on behalf of the subscribers),
H. R. FALKINER, President N.T.C.C. W.
H. DODERY, Treasurer
GEO. SCOTT, Secretary
'What obsequiousness, Holas!'
'The response was worse, Blue. Madam wrote it.
Quamby,

9th July, 1880.

To H. R. Falkiner, Esq., W. H. Dodery, Esq., George Scott, Esq., and the gentlemen who have subscribed to the beautiful and valuable testimonial of your good wishes presented to Mrs. Sadleir and myself this day.

Gentlemen,

We wish to convey to you our most sincere thanks for the expression of your feeling towards us. It did not require such a valuable memento to remind us of the many happy days that the N.T.C. Club and attendants provided for us and for our friends, and we felt favoured by your presence. The few little attentions to the public by Mrs Sadleir and myself were simply what would have been offered by any Tasmanian who had hares and land to course them on. I would wish to mention that the Messrs Phelps desired me to offer the use of the Quamby estate to the N.T.C. Club to hold their meetings on, and show them whatever attention I desired, and to preserve the hares for their use (until a more suitable place offers); and this with great difficulty I have done.

All those who have attended your meetings we were delighted to see enjoyed themselves, and they never gave the least annoyance or did injury to the estate. Again allow us to repeat our most sincere thanks for your kindly wishes. Your Club and gatherings shall be always welcome at Quamby, and we hope to secure plenty of hares to run off large matches that will gather the whole colony to us. Allow me to subscribe myself on behalf of Mrs Sadleir and family,

Yours sincerely,

N. SADLEIR.

And you will notice in the letter, Blue, Madam mentioned *on behalf of Mrs Sadleir and family.* By 1880 we had five children: Eliza Georgina, who came with us from Albemarle, James Phelps, our first son, Mary, Richard De Vere and Georgina Hunt. I think the air agreed with us. Five in six years!'

'And you were busy with other things too, Holas. Commercial divinity! Perhaps you balanced your heavenly accounts by managing St Margaret's finances.'

'You do have an impudent irreverence, Blue.

In the years after moving our household from Albemarle to Quamby, John and Joe returned to Ireland and married there. John's wife came to Albemarle for a short period before they returned to their home at Willowbank near Limerick. They stayed with us here at Quamby but Mrs Phelps was not amenable to the colonies. I think John wanted to install her at Albemarle and she liked it about as well as Mrs Sadleir did. They went back to Ireland as soon as they left us at Quamby.

Joe's wife never came to Australia and so Joe used to go backwards and forwards.

John and Joe both had sons born in Ireland. John's boy died early and Joe's boy lived only into his mid-twenties. I was sorry not to have met them. In many ways John and Joseph were like parents to me and their sons would have been like younger brothers.

The greatest Phelps' legacy from their British residency were several importations of fine breeding animals. We unloaded grand examples of Southdown and Lincoln rams, and Hereford bulls at the Launceston docks, but two entire horses had the most celebrated arrival. They were both heavy horses, a Clydesdale and a Lincoln, and we sold their services to owners of mares in the whole of Tasmania. We sent them out on the railway and invited farmers to bring their mares for service at railway sidings for a fee. It worked well.

I went back to Albemarle twice in the first year at Quamby to advise John Phelps who was living there, but after that, I usually went once annually for up to two months and I was able to combine those trips with a visit to Bingara or Wilcannia to see Fitzherbert Brooke. Frederick Vandeleur, a relative of John's wife, replaced John as manager of Albemarle and things went smoothly with my supervision of him from Quamby. We wrote to each other at least weekly, and we used wires for more urgent matters. We continued to improve Albemarle and stock increased. We pumped water for the homestead gardens from the river with a new steam engine driving the pump, and we put in dams and wells for stock water away from the river.

'Did your mining experience help with the well sinking, Holas?'

'It did and it didn't, Blue. A couple of coves had come up to Albemarle directly from Clunes looking for contract work in well sinking because of their background in mining. We were negotiating prices and I said, "The Albemarle rate is based on two yards a day per man." The senior man said, "Have you supermen on Albemarle, Sadleir? That is a man-killing rate." I was a bit younger then, Blue, so I said, "Watch this, my good sirs, boil a billy. I'll be looking for a drink of tea in about half an hour after I've moved half a yard." And I shed my coat and did it, drank the tea and said "There, you have an Albemarle bonus given in – three and a half yards to go. You should finish it by dinner time and have plenty of time after that to set up your camp."

'The word spread. Whenever I started contractors on a new well they would say, "Sadleir, can you show us the Albemarle Superman style."'

'Well what did you do?'

'Usually I just grinned and said "To your work gentlemen – or something equally evasive."'

The staff supported Madam well in the household at Quamby and so I could leave without being anxious about her and the children's welfare. Sams managed Quamby when I was away. I had joined the Melbourne Club and I used it frequently on trips to Melbourne for business and to assess livestock markets, or to stay overnight on the way to or from Albemarle, Boondarra and Bingara, and I was able to expand my circle of contacts of like-minded investors in pastoralism. I enjoyed the sport and the race meetings as well and I saw Richard and his wife Eliza regularly, and John and Marshal and their families when they were able to come to town.

I encountered Redmond Barry at the Melbourne club several times. He was Sir Redmond by then. He remembered me fresh off the boat from Ireland. "Nicholas Sadleir. I expect you still regret missing a career at the bar." But he was joking. He knew of my reputation.

Fitzherbert Brooke and I dissolved our partnership on the Paroo. John and Joe Phelps were pleased and they rewarded me for my loyalty to them by doubling my salary to £2000 and asking me to supervise the farm manager at their farm at Laverton near Melbourne.

'And you persuaded two hard-headed Irish Quakers to buy a mansion in northern Tasmania because Great-Grandmother was unhappy living in the West Darling, Holas.'

'A question or an assertion Blue?'

'What do you think?'

'Let us leave it as an assertion. John and Joe did very well out of it. In 1882 I had Quamby valued at more than £90,000[1] and we realised more than that when we sold the individual farms. Perhaps it is fairer to say that all parties were contented.'

'Modest of you, Holas.'

'Thank you, Blue.'

---

1. £90,000 in 1882 = $A 22,500,000 in 2012

Nicholas Clarke Sadleir, aged 47.

Nicholas reminisced about entertaining two of his brothers at Quamby. 'In 1882 my mother sent me a strange letter from Tipperary:

Brookville House
Tipperary
14 March 1882

My Dear Nicholas,
To begin I must thank you and your darling wife Anna for the pretty photographs of your children you sent me. Photography is a wonderful thing and we must give thanks for it.
We are approaching a grand spring and our farmers look

forward to renewed activity and prosperity as the days grow longer and the weather gets warmer. James is installed here with his new wife Emma and they help to fill what was previously an empty nest. Most of you, my darling children, have left for Australian colonies. Have you news of Helena? I have not had a word from her for some time.

All seems normal here but I fear it is not. The Church of Ireland has already been disestablished. Insurgency in this beautiful country seems ever present but its evil tentacles seem to be spreading to London – Irish emigrants from America who call themselves Fenians are dynamiting buildings in London as a means, they say, of persuading Prime Minister Mr Gladstone to restore a separate Irish Parliament. They disrupt ordinary life with a method they call Terrorism. So far, their activities remain in London but I fear they may cross the Irish Sea to wreak their mischief here.

I am concerned about this so-called Terrorism emanating from the Irish in the colonies and I worry particularly about John and his role in the Victoria police force in struggling against what seems to be a similar development in Victoria. We have recent news of the so-called Kelly gang and their martyrdom at a place in Victoria called Glenrowan.[2] I have written to John and to Marshal seeking their opinion about this burgeoning colonial menace but they have not responded. I imagine perhaps they dismiss my entreaties as the imaginings of a silly old woman, but please Nicholas, I beg you to report to me on the perils to the British Empire of this new Australian Fenian development and of the hazards to John and his family's safety in attempting to dismantle it – and Marshal, as an attorney, may be menaced as well. I believe he resides in the district harbouring this turmoil. But equally importantly, I beg of you to use your influence as an honorary magistrate and gentlemen to see this evil menace is countered and cancelled.

I beg of you Nicholas, respond quickly.
Your loving mother,
Elizabeth Sadleir.

---

2. This was the site of the famous *Last Stand* of the Kelly gang. Three members of the gang died and Ned Kelly was arrested.

I asked Anna to read it and explain it.

"Your mother is clearly disturbed, Nicholas."

"You mean not of sound mind?"

"No. From the construction of the letter I do not judge her to be feeble-minded." And she looked at me and ducked her head. "I think she is genuinely concerned. She is appealing to you to correct matters."

"But what does that mean, my dear Anna? I have never heard the expression *Terrorism*. Does Mater believe supporters of the Kelly gang are plotting to blow up a railway station in Melbourne, and that I should rush to prevent it? Are you concerned Anna? Do you really believe the good horse-stealing folk of Greta in Victoria are conspiring with American Fenians in New York?"

"No, my dear Nicholas, but perhaps your mother does."

I left the letter in my writing desk. I talked about it later with John and Marshal who had flitted across Bass Strait on a steamer from Port Melbourne to join me and the Northern Tasmanian Coursing Club in two days of fun at Quamby judging the speed and proficiency of greyhounds and wagering on their combination. Neither were accompanied by their families (Marshal had come from Mansfield alone to join John in Melbourne) and they had the spring and swagger of single men. John looked especially unbound. He had just been through an ordeal of a police commission enquiring into conduct concerning the detection and capture of the Kelly Gang and he had fought strenuously to defend himself and his colleagues. He had prevailed but he was exhausted and looked forward to two days of coursing. He had brought a greyhound bitch with a litter of pups with him as gift for me.

Richard had not been able to come. He was encumbered with the conduct of an accouchement hospital, and when he complained, we reminded him of the encumbrances of the profits as well. He had become a rather sensible and dull fellow.

At the end of the second day we were toying with our brandy after a fine dinner. Madam had left to attend to the children and I raised mother's letter.

"Gentlemen, have you received a letter from our mother fearing a connection with the Kelly gang here and the American Fenians who are dynamiting London?"

They said they had. "But the Kelly gang are gone." John said. "Ned has been hanged. Dan Kelly, Steve Hart and Joe Byrne perished in the hotel fire we lit at Glenrowan – or they died of their wounds before the fire overwhelmed them. We were never certain."

Marshal intervened, "Concerning our letters from Mater. She is obviously less bold than we remember her. I think she is imputing more in the activities of the Kelly gang than perhaps we might concede. I believe some of the Irish press have painted the Kelly gang as activists in a worldwide movement encouraging the emancipation (and that is probably an incorrect term) of Ireland. I think that has alarmed Mater, she remembers some of the Tipperary bloodshed before the potato famine – remember Uncle Patrick – and, as a mother, she worries about us."

I passed them my letter from my mother. "How does this differ from the letters you received?'

"Well I don't think we received the lofty flattery that Mater heaped on you, Nicholas – honorary magistrate indeed! And she was not to know that the judiciary function applied to New South Wales and had no currency in the colony of Victoria, or that you are a resident of the colony of Tasmania without Kellys, but even so, you should lead us in formulating a plan to defeat this evil menace."

"All right, gentlemen. You have the better of me. I would love to see the letters you received, but clearly I may not." I passed them the brandy decanter and poured more coffee. "John, I remain curious. I am sorry to raise this. I'm sure the whole Kelly business and the subsequent police commission has exhausted you, but was there anything political in the actions of the Kelly gang?"

"A difficult question, Nicholas. As you know, I spent months hunting them after they had murdered our comrade. The Kellys had a great deal of support from members of their class and that support was mixed with genuine affection and with collaboration in the distribution of the proceeds of crime. And the gang was betrayed – so the motives were mixed. Some of the Kelly gang supporters were able to quote the exploits of Irish patriots but I think many of them were just struggling to make a living and stealing livestock was just one of the ways. Certainly most of the small farmers envied the wealth of the larger landholders. But..."

Marshal interrupted. "I think there is still a marked divide, John. It is mostly in people's heads, but it is Protestant versus Catholic, landed versus landless, just as it was in our old country. But if our Roman Catholic countrymen bothered to look, when they become free men in these colonies (and remember many of the first generation came as convicts) they have far more rights here than they had at home.

John resumed. "I agree, Marshal. I think if there had been someone like you, or a useful priest, or even sufficient police constables to maintain law and order to show say, for example, the Kelly family, that they did have rights, they would be treated fairly, and that the rule of law applied equally to everyone, then things may have been different. I suppose it is natural that I would, but I think deterrent to crime is the best defence against it. Had there been sufficient fair constables in the Greta district when Ned and the others started their small affairs in stealing horses, they would have been dissuaded. I've listened to stories about the Kellys and I think some of the early encounters the family had were not with the best members of the Victoria police and they were treated grimly which soured them. But I always thought of them as ordinary criminals (and remember, I did not know them from the beginning, I came to the investigation late). And, in the end, I thought of them as murderers of a policeman. I think they elevated themselves to appear as revolutionaries later on. It may have been a mixture of self-delusion and a sort of madness that came with the belief that they were beyond arrest and could continue being merry bushrangers forever, with no safe home or haven. Going on from that – the armour they wore – attacking us by hoping to derail a trainload of police and capturing people in the Glenrowan Hotel led to a tragedy that should never have been."[3]

---

3. He wrote about it in his book: Sadleir, J. 1913. *Recollections of a Victorian Police Officer, 1913,* George Robertson & Company Propy Ltd, Melbourne, Sydney, Adelaide, Brisbane and London.

Marshal said, "The gang did have a lot of popular support, John. Things were nasty in the district for a while after Ned was hanged."

"Oh yes Marshal," John responded. "One of the leaders of an inner circle of sympathisers approached me and told me they intended to go on with things, but he paused a little in his bravado when I pointed out to him what happened to the Kelly gang, and that he could not eliminate a total police force – perhaps a constable here or there may be shot – but a police force would go on forever. He became more reasonable at the end of the interview. He ended it by asking that those of the Kelly circle who had taken up land not be dispossessed. I was able to promise him that nobody who kept the law would be interfered with, but I went on to say that no further selections would be allowed to doubtful characters. We formed an important truce.

Marshal nodded. "I still spend much of my time defending my people against the police.'

And John said, "And so you should Marshal. That's your calling. Police need to be kept to the straight path. The difficulty in this police commission I've just been through was that many of the police were being treated in the way that you say many of your clients are. There were all kinds of uncorroborated evidence and it was a kind of witch-hunt without proper cross-examination. I had to work very hard to defend colleagues and myself.

May we talk about something else? No one is going to dynamite buildings in Melbourne."

I said, "Well what do we say to Mater?"

Marshal said, "I shall write to her on behalf of us all. Richard has a letter as well that he doesn't know how to respond to."

Marshal sent us copies of the letter:

Mansfield
Victoria
1st of June 1882

My Dear Mother,

Thank you for your last letter sending us your good wishes. I am pleased to report that we are thriving at Mansfield, two of my sons are working with their Uncle Nicholas either at Albemarle or on a station he shares further north in Queensland called Bingara and all the children are well.

Since I received your letter, I seriously considered the grave concerns contained therein and I took the opportunity of discussing it with Nicholas and John during a recent visit to Quamby. Nicholas invited us to coursing matches there. John and I stayed for two nights and it was a grand family reunion. It was a pity Richard and Helena were not present. Richard could not absent himself from his hospital duties and Helena did not respond to Nicholas's invitation. We believe Helena is in a remote district of Western Australia and the mails are slow and unreliable. John has made enquiries of the Western Australia authorities in search of Helena's exact location and he hopes for a favourable response soon.

To the matter you seek our advice about – colonial conspiracies of insurrection. Since you told me of the North American-based Fenians and their work with dynamite in London, I have found copies of English newspapers to read brief reports of the carnage left after the explosions and attributed to Fenians. It is indeed a grave threat to the good order of things and we can only hope that Prime Minister Gladstone will remain steadfast in opposing this extortion. We do not believe there is any connection between the North American Fenians who are active in London and the Kelly gang who have recently been defeated at Glenrowan in Victoria. The Kelly Gang were criminals, not political revolutionaries, and that is the way most people here remember them.

As for our safety, Mater, there is no hazard from revolutionary people or tendencies. Richard continues to thrive as a surgeon. Nicholas delights in his duties as a steward to your cousins, the brothers Phelps, with their large agricultural and pastoral holdings in Tasmania, Victoria and New South Wales, as well as developing enterprises with a partner in Queensland. John now has major responsibilities for the police force in Melbourne and I continue with my modest practice of law in the beautiful Mansfield district of Victoria. I expect we will report on Helena's progress soon.

As a rule, we delight in the company of happy, just and healthy people who thrive in these new colonies.

Your loving son,

Marshal Sadleir.

Marshal made a fine effort to mollify our mother. But I doubt she was convinced. I smiled when I read in a newspaper several years later that Prime Minister Gladstone had had agreed to an Irish Parliament and planned to introduce a bill to the House of Commons. Perhaps my mother was right about the influence of Fenian dynamiters from New York but I hoped brother Marshal had convinced her that the Kelly gang were not responsible.

A lot more came from that conversation in the Quamby dining room. John told us of the theory he developed using a modern term, *paranoia*, to describe the unusual fear the Kelly gang had for the members of the Queensland Native Mounted Police he had recruited to track them down. "It seemed to change them. They became more fearful and reckless at the same time. When Ned was lying wounded after we had captured him at Glenrowan, a railway fireman stooped over him to offer him comfort and Ned screamed *get away you black bastard*, with rolling eyes. He was terrified. The railwayman's face was blackened with soot. Ned thought he was a black tracker."

Queensland Mounted Native Police Constables with Victoria Police Officers. John Sadleir is second from right with the bowler hat.

Marshal reminded me that Redmond Barry, who had condemned Ned Kelly to death and died soon after Ned hanged, had tried to persuade me to a legal calling when I was newly arrived in Victoria. "It could even have been you on the bench, Nicholas." And we talked about how John, the arresting officer, and Ned's father had come from Tipperary, and Barry from neighbouring County Cork. And we compared our differing circumstances. Ned's father, the convict Roman Catholic, arrived in the colonies from Tipperary for stealing pigs. We, the Protestants, came as free men looking for gold or new vocations in a developing paradise.[4] We toasted our luck.

But we could not find Helena. John wrote me a short note soon after he returned from Quamby.

Russell Street station
Victoria Police
September 11, 1881

Dear Nicholas,

I have had letters and wires from the Western Australian police and they have no trace of a Helena Sadleir in any of their districts. They have searched ships' passenger lists and they believe that she never arrived in Western Australia.

Since then I have formed another line of enquiry. A colleague from the South Australian colonial service believes that somebody answering her description may have worked in a charitable school in the Port Adelaide district some years ago.

Please keep this between us. I will seek more information from the South Australian colonial police. Please be patient.

Yours,
John.

Great-Grandmother talked about her children at Quamby.

'I loved having children, Robert. And I loved teaching them. For the first ones, Elsie, Jim and Mary, I taught them myself in a special schoolroom I had set up in the upper storey away from the bedrooms in Quamby. When they got too many some went off to school every day at the Hagley School; but I loved supplementing

---

4. It was said Australian colonists had the highest standard of living in the world in the 1880s.

their learning and I kept the schoolroom as a schoolroom and I encouraged the children to play there and play at being in school, one being the teacher and the others pupils. Sometimes they rehearsed and performed special pieces of theatre for Nicholas and me there. We loved it. Nicholas was an especially loving father.

Nicholas was away for some of the time but I didn't miss him terribly in the way I used at Albemarle. He was not away for long periods usually and I had the children, and support of a loyal household of capable servants and drivers. Nicholas always sent wires asking how we were and telling us where he was whenever he could get to a Telegraph Station.

Scarlatina nearly laid us low in 1882. I was beside myself, but Nicholas was at home. He was sometimes firmer with servants than I could be and we were able to get the doctor in the middle of the night because Nicholas sent out a buggy and driver to fetch him from Launceston several times, and twice in the middle of the night. Elsie, Georgie and Jim were all frighteningly sick. I had just come back from Georgetown. I had taken the children to a seaside holiday up the Tamar, by myself, but with some servants – Nicholas had remained at Quamby, it was a busy time on the farm. We had been yachting at Beaconsfield and Beauty Point (and I fear they may have caught the infection there) and soon after we returned home, down they went – terrible temperatures, sore throats and as weak and floppy as rag dolls. I nursed them night and day and the servants were wonderful. Eventually they got better, but I worried that Jim recovered slowly and he was sick and ill-thrifty for a long time. We went nowhere for two weeks. It was school holidays and the children who were not sick, Richard and Mary, had a miserable time as well. It was quite infectious. Several of the servants got it. I was pregnant and I feared for my own health and the health of the baby, but we all got out of it all right except Jim.'

Nicholas spoke. 'I took Jim to Melbourne to see Richard months afterwards. Richard told me it was a dangerous disease, but he thought that Jim would get stronger. He was right. But it took a long time and Jim never had much stamina. He was never keen on games.'

He went on. 'Quamby wasn't entirely a children's paradise. We left two behind in the churchyard at St Mary's, Hagley. Dick was a lovely handsome boy and he died of diphtheria when he was nine. It was a horrible illness and it broke our hearts. William died as a baby soon after he was born, so that was less sad. He didn't live long enough for us to get attached to him. We don't know why he died. Dr Payne said he thought it might have been because he had a weak heart.

But I suppose where there is life there is death and it is wonderful that we had 15 children born.'

I read from a list of the children.

'Eliza Georgina, born 1875. Lost at sea in 1905 on a voyage from Adelaide to Melbourne.

James Phelps, born 1876. Died in 1931 at Yorketown, South Australia.

Mary, born 1877. Died at Adelaide, South Australia 1959.

Richard De Vere, born 1879. Died 1888 at Quamby.

Georgina Hunt, born 1880. Died 1912 at Murray Bridge, South Australia.

Nicholas Sturgess, born 1881. Died 1896 at Glenelg, South Australia.

Robert Hunt, born 1882. Died 1961 at Comodoro Rivadavia, Argentina.

De Vere Ralph, born 1884. Died 1946 at Cordoba, Argentina.

William, born 1885. Died 1885 at Quamby.

Charles Hare, born 1886. Died 1959 at Yalgoo, Western Australia.

Angela Margaret, born 1888. Died 1970 at Adelaide, South Australia.

Aubrey Toler, born 1889. Died 1946 at Cottesloe, Western Australia.

Alicia Grace, born 1891. Died 1948 at Adelaide, South Australia.'

John Raimond, born 1893. Died 1918 at Grantham, England.

Kathleen Ross, born 1895. Died 1977 at Adelaide, South Australia.

The last four came to you in South Australia I think.'

The two started talking at once.

'Yes your list is correct.'

'They are all dead already?'

'Perhaps we can start with a story about each one?'

'Yes,' Great-Grandmother said. 'Let us begin with Eliza Georgina – Elsie. And unlike you Nicholas, I know her life from start to finish. She died a year to the day after you did Nicholas. Elsie was the only one from Albemarle, and she was the first, the one to make mistakes on, the one to spoil. And everyone did. When she came to Quamby there were about 20 people hanging on her every whim. She could be a handful. When the other children turned up, especially Jim, she would run away and sulk if she thought she wasn't getting enough of my undivided attention – and obviously she wasn't, I had new babies to look after. Do you remember Nicholas, once she ran away and we found her hiding in the corner of the slaughterhouse – she had made a kind of tent for herself using drying sheepskins and furnished with cups and saucers and so on?'

'Yes I remember, my dear. We stopped ploughing and carting wood that day to have enough men to go out searching. We didn't find her. She got flushed out of her cubby house when the butcher drove in a big bullock to settle that evening for slaughter the following morning. Elsie had the sense to get out of it and she ran straight into my arms as I was walking past. The bullock found her! She sobbed as I carried her all the way back to you.'

'Yes, the darling Elsie. Once she got over the shock of the other children arriving she became an assistant mother by the time Georgina was born.'

'Assistant mother! More like Sergeant Major. She had the others toddling off after her to collect tadpoles in the pond in the garden. And then she kept all the tadpoles and wouldn't give any to the others. What lovely fights they used to have.'

'Elsie did well at school. She was a joy to teach and it encouraged me to keep the schoolroom going and for her to encourage the others to enjoy learning as much as she did. She was probably our best scholar and she loved taking charge of the plays the children used to put on the schoolroom.

'What about Toler?'

'Oh yes, Toler, I forgot about him and his scholarships, but that was much later, he was born and grew up in Adelaide and I taught the children less then. We are getting ahead of ourselves, but even so, Elsie spoke the best French.

She had an early death. Elsie nursed Nicholas at Albemarle with her brother Robert when Nicholas died. She seemed wounded when her father died. She hosted the whole family for the funeral (we came up on the train and Mr Kidman put on special coaches for us from Broken Hill to Menindee) and she did that very capably, but when she came home with me to the Grange she appeared to lose all passion for life. It was tragic. She was listless for a year. My brother in law John invited me in a letter from Melbourne to visit him with Eliza for a holiday and to talk about his brother Nicholas to see if she could improve her feelings by listening to stories about her father. Our family doctor had been treating her for nervous prostration without effect.'

Great-Grandmother paused in silence. When she resumed she spoke softly. 'That journey to Melbourne was tragic. Elsie went missing from the steamer Bombala out of Port Adelaide bound for Melbourne. The police in Melbourne suggested she might have struggled out of a porthole in our cabin, fallen into the sea and drowned. What a waste, our firstborn gone at thirty in such despair! We commemorated hers and her father Nicholas' life on his gravestone at Menindee. She went missing at sea a year to the day after he died.'

'I came back to Adelaide alone on the train. It was awful. John had been kind and sympathetic and he helped with the dilemma I had about what to tell the children. "Tell them the truth, Anna"– but what was the truth – in the end I simply told the family "Elsie disappeared at sea. The police do not suspect foul play. Sadly I think she did away with herself."'

'Terrible for you, Great-Grandmother.'

'Saying it like that was a relief, Robert, and I had very little time for sadness. The children said nothing and they asked no questions. DeVere came up with a suggestion to commemorate Elsie and Nicholas's life on Nicholas's tombstone at Menindee, and he commissioned the gravestone and had it freighted to Menindee via river steamer and erected there.

Nicholas and Eliza's grave at Menindie

There was a long silence. Eventually I asked about James Phelps. 'His first name James was obviously for his grandfather back at Brookville House, but where did Phelps come from?'

Great-Grandmother answered. 'I think it came from both of us as a way of saying thank you to the Phelps brothers for the way they looked after as both from the minute we were married. Nicholas was somewhat abashed about it, he thought that people would think that we were trying to curry favour with our employers who also happened to be cousins, but no one knew James's second name, and if they did, I doubt they gave it a minute's thought. I think Nicholas was more exploitive of James than he was of the Phelps. Dear little James was a pastoralist by the time he was seven.'

'My dear,' Nicholas interrupted, 'you make it sound as if I sent him down a coal mine as an infant. I was simply investing for his future in a partnership with Richard Wilson and Frank Brush in a place called Boondarra. I can remember arranging the partnership in Melbourne the day after brother Richard's funeral in Toorak. He died too young but he left a grieving widow only – no children.'

I Interrupted. 'I went looking for Boondarra a couple of years ago Holas. Someone directed me from Mossgiel east along the Hillston Road. I had just driven there from Victoria Lake.'

'You were out for a couple of days, Blue?'

'No, Holas, probably two hours from the Victoria Lake woolshed. I was in a motor car. Not on horseback or in a buggy like you.'

'But there was no Victoria Lake woolshed. We used to walk all the sheep into the river for shearing at Albemarle – at the homestead. We had 80 shearing stands there in the early days – two shearing boards, one on each side of the shed. Later on, when we sold some country, I reduced it to one shearing board and had a mechanical shearing plant with 40 stands driven with a steam engine.'

'There is a shearing shed at Victoria Lake now, Holas. This will interest you: Long after you died, the Irish owners lost the old homestead block to closer settlement and they were left with the Victoria Lake block. They took the Albemarle name with them to Victoria Lake. The old Albemarle homestead block is called Windalle and Victoria Lake is still called Albemarle. What perhaps is worse, is that they dismantled the woolshed and shearers' quarters from the riverbank you knew and had all the galvanised iron sheets carried out to Victoria Lake to put up new buildings there.'

Great-Grandmother interrupted. 'I dislike the silly names you use for each other and this does not have pertinence to James Phelps Sadleir's pastoral holdings at Boondarra as a seven-year-old.'

'I agree, madam. Great-grandsons can be a pest.' He paused. 'How did things look at Boondarra, Blue?'

'It was a wonderful season, Holas, green feed everywhere, but a lot of it was ephemeral stuff, annuals, some grasses but a lot of puffy herbs full of water and not much nutrition. Any saltbush or bluebush that may have been there when you had the country is gone. I doubt it is country that will carry livestock for long periods in a poor season.

'Any buildings, Blue?'

'I got to where the old woolshed had been. Somebody directed me there from Mossgiel. She was a drover's wife who lived in the old Mossgiel Post Office.

'There was no post office when we were there.'

'I know,' I said. 'But there was a station called Mossgiel. I think it became a town after your time. It even had a hospital. There is nothing there now.

And I suppose I could say the same about the Boondarra woolshed. It was no more when I saw it – just a pile of fence posts, timber, wire, roofing iron and dirt. It looked as if someone had demolished it quite recently. The woman at Mossgiel said that Boondarra was part of another place now and the woolshed was the only building remaining from your time there. She did not know if there had ever been a homestead.'

Boondarra woolshed demolished

'Yes, it wasn't a large run. We never had more than 25,000 sheep on it. It was a bit more than 120,000 acres and we had it for only seven years, and the manager and men lived in temporary huts. We dissolved that partnership in 1890 – we were in Adelaide then.'

'So little Jim was only 14 when he became landless?'

'Yes I believe Nicholas never gave shares in stations to any more of his children after that,' Great-Grandmother said.

'Well there was too much country, or too many children. If you add Quamby at Cloncurry to Mingara out on the Queensland and Northern Territory border on the Barkly tablelands we had more country than Albemarle – more than a million acres I think, about the size of Tipperary county, and I changed partners with those runs. Frank Brush of Melbourne bought Jack Phelps' share. Some of my nephews were growing up and they may have been interested. Ernest, John's boy, went with me to Quamby, the one at Cloncurry, and was there for several years.'

Great-Grandmother interrupted. 'And he, poor boy, came home a wreck, and Marshal's boys, James and Stephen, spent time at Albemarle and Bingara and they disliked it too. And worse still, Richard, Marshal's son died at Albemarle in an accident with horses. You may not remember, dear Nicholas, but I think I helped you form the view that a life of pastoralism was not necessarily the best choice for our sons.'

Holas remained silent.

'Mary was a lovely baby,' Great-Grandmother said.

'All mothers say babies are lovely,' I said.

'Well even so, she was. She was a lovely girl and young woman. She supplanted Elsie as the assistant mother of the family – Elsie remained the leader of adventures, but Mary was the mother. She was steady and dependable. Whenever she decided on something to do, it was sensible.

'Yes I met her when I was a little boy when she lived at the Grange. I think I told you about the house, Great-Grandmother. She was very much the lady of the household then but only Kathleen was there with her when I met her first, and Angela joined them from Argentina later.'

'Why was de Vere called de Vere?'

Nicholas cleared his throat. 'I think that is a question best answered by Madam.'

'Well he was called de Vere Ralph, and Ralph comes from your family, Nicholas. He was a famous ancestor who helped Henry VIII and his daughter Elizabeth to control Scotland.'

'Yes but poor little de Vere was stuck with his odd name. We never called him Ralph and I'm sure he had no idea who Ralph Sadleir was. I think you should answer Robert's question more straightforwardly, Madam.'

'Well then,' and she paused. 'Robert, I think you know my mother Georgina Hunt was directly descended from the Earls of Oxford. My grandfather, John Hunt, formed part of the Irish branch who came from Jane de Vere who was a daughter of an Earl of Oxford. Jane de Vere married a Hunt and a son John went with the English armies to Ireland and got land in Wicklow and Limerick just as Thomas Sadleir did in Tipperary. Vere or de Vere was a common name for Hunt men throughout the generations. I never met any of my Irish Hunt relatives when I was in England as a girl, and I regretted that, but I thought that using the name would acknowledge my and my children's link with the Hunts of Ireland and the de Vere's of Oxford. As you know, I never knew my mother, and my father had a sketchy knowledge of my mother's family, so I wrote to members of the family who called themselves Hunt or De Vere in various parts of Ireland and they sent me pedigrees and family trees. It was most interesting. Have you heard of Aubrey de Vere?'

'The Irish poet?'

'The same. He was Sir Aubrey de Vere, he and his brother Sir Stephen de Vere were baronets of Limerick. They were my relatives. At some stage their family changed their surname from Hunt to De Vere.' Somewhat defiantly, she added. 'It was important that my children knew about their ancestors.'

There was a pause. I asked, 'Tell me about leaving Quamby. Why did you go to Adelaide? I expect you were sorry to go.'

Nicholas answered, 'Yes it was sad for Madam and the children and I loved my life here. It was almost perfect. We've talked about the coursing I started in northern Tasmania, we had horse race meetings as well at Carrick and Deloraine, and Quamby regularly played a team from the tennis club at Westbury at the courts we maintained at Quamby. We had many visitors. Anna's father came regularly from Geelong, the Bishop and the Governor of Tasmania were regular stayers, members of my family visited when they could, Fitzherbert Brooke was able to come down from Queensland on several occasions – the children loved him, he was a sort of an uncle. And we looked after Jack Phelps and his sisters for periods when their father Robert became a widower. We all wanted to stay, but it wasn't to be.

It started when John Phelps died in Italy on a holiday from Ireland. He and Joe were the sole partners in Albemarle and Quamby and they had a farm at Laverton. I superintended that and I looked after various other interests they had in land in Melbourne. It took me two years to value the holdings and to help to divide the assets fairly. John's executors were Joe, who spent most of his time in Ireland after John died, brother Robert, the winegrower from Albury, and myself. Joe owned half the assets and worked towards making a fair distribution to John's heirs. We talked of selling Albemarle (the agents advertised it for sale at one stage but we cancelled it). In the end, Quamby went with blocks of land in Toorak and the farm at Laverton, but we sold it far more successfully than the way in which Lady Dry did when John and Joe Phelps bought it. We sold the tenanted farms individually, we subdivided some of the farms we had resumed for grazing to sell them more favourably, and it was grand to see tenants buying farms they leased. It took a long time to finish things. We were not able to sell all the farms at the first auction, but our lovely home went in the first round with about 600 acres surrounding it, and so we moved to Adelaide.

I came to Tasmania several times for four years after we left, but I had some good people in Hobart and Launceston who finalised affairs for the estate. By then I was busy with Quamby and Mingara in Queensland and Joe kept Albemarle for me to manage – Jack Phelps, Robert's son was in charge there. Now all the journeys were from Adelaide – railway from Adelaide to Broken Hill and coach to Menindee, or railway from Adelaide to Marree to meet cattle walked down from Queensland via Birdsville, or steamers to Townsville and coaches inland to Cloncurry and beyond. No more port and cigars at the Melbourne club and race days at Flemington.'

'And that was no bad thing, Nicholas.' Great-Grandmother resumed. 'The children were angry and sad and frightened. It was a perfect and privileged life for them. Jim had started his secondary schooling at Launceston Grammar. Mary and Elsie were with a private tutor of music and French in Launceston. Georgie, Nicholas and Robert were in school with me or at the Hagley school and the others – De Vere, Charles and Angela were toddlers or babies looked after by me with the help of staff in that wonderful household.'

Thoughts turned to Queensland. I, the great-grandson, was in Cloncurry in 2011, contrasting Great-Grandmother's Tasmanian gentility with Nicholas' Queensland nearby pastoral investments. It was still a rough mining town in northern Queensland. There were old pubs looking as they did when they were built in the late 19th century – Nicholas possibly knew them, and it was Friday. Miners were bustling in for the weekend.

'Quamby. There is nothing there but a pub. A railway ran there years ago but that's gone and I guess the township went at least 50 years ago.' Gail at the museum said. She had helped me find Quamby and Timbaroo marked as pastoral leases on a map she had from about 1930. These were the two runs Nicholas had taken up with Jack Phelps and Daniel McDonald in 1882. A town called Quamby sat on the map too, abutting the runs.

I drove to Quamby, over the Cloncurry River and for about 20 miles north-west through stunted scrub on the Burke Development Road.

Quamby Hotel, north of Cloncurrry, Queensland.

I arrived at the Quamby pub. It was the only building there. There was no sign of a township. The publican said it disappeared years ago. When I asked what Quamby meant she said, 'I reckon it means water or meeting place in the local Aboriginal dialect.'

I paused for a bit and grinned (I wished I'd had a dollar for every Aboriginal place name said to mean water or meeting place – but I kept my tongue).

'I've got another theory. I'm sure you've heard of Quamby Estate in northern Tasmania, your pub having the same name as it and everything?'

'Yair.' I could see I had her attention.

'Well my great-grandfather lived at Quamby in Tasmania at the same time as he was establishing a station called Quamby with partners here.'

'Yes there's still a block called Quamby on a neighbouring station out the back of here,' and she gave me a phone number.

I promised to give her more detail via e-mail about Nicholas Sadleir and the coincidence of two Quambys. I told her I was pretty certain that my story was true, but she didn't look as if she wanted to be convinced.

'I like the way you've done the pub up, what's your main drawcard?'

'This is a party pub.'

'Yeah but what does that mean?'

'Cold beer and hot girls. We have some good weekends. We've got accommodation too – miners' accommodation. There is a working zinc mine within half an hour and a lot more new prospects are opening up. We are looking at taking up another pub licence at another mining town site about half an hour away.'

It was early afternoon. I was the only customer but there was lots of room for many more. The pub was rustically decorated (there was the body of an old truck at the front door, and other bits of harness and equipment decorated the walls) and it felt comfortable.

I said before I left, 'Yes, I think this would be an ideal party pub. It doesn't match the Tasmanian Quamby for elegance but I reckon it would be more fun.'

It was hard to tell if she thought it was a compliment.

I found no sign of a town in the scrub behind the pub, let alone an old station homestead – all gone years ago.

Holas chuckled about the conversation I'd had at the pub. 'Aboriginal for meeting place indeed! History does not seem to last long, Blue.'

'How did you stock the runs up here, Holas?'

'Daniel McDonald came right down to Hay in New South Wales to collect horses we bought for Mingara in 1882. He had some blackfellows with him from Cloncurry and they took the horses to Bingara, picked up some of my cattle as the last part of my settlement in the dissolution of my partnership with Fitzherbert Brooke, and we had 1000 cows and about 50 horses to start Mingara.'

'How long did it take him to get to Mingara?'

'I think he was on the road for six months at least, Blue. It was something less than 2000 miles. From memory, I think he went with the horses from Hay to Hungerford, picked up the cattle from Fitzherbert a few days further north and then started the long plod out to the Barkly Tablelands country.'

'And when you started Quamby near Cloncurry you transferred stock from Mingara?'

'Exactly, although there were a few cattle there, and we ran sheep on Quamby as well.'

'How did you staff the runs, Holas?'

'Mostly with blackfellows, Blue, but they got troublesome and in the end many were wiped out, so we relied mostly on people from further south. I was able to send up some coves from Albemarle. One of them was my nephew Ernest, John's boy. He worked on both places for several years but he came south after the partnership dissolved and he ended up in Western Australia.'

'How did the Aboriginal people get troublesome, Holas?'

'They murdered settlers and the Queensland colonial government had to take special measures. There was a contingent of native police in the district but the killings went on so the government sent up a new Sub-Inspector.'

I interrupted. 'I've looked him up, Holas. Frederick Urquhart was his name. He wrote poetry. He seemed a fierce, strange man. Here is a poem he wrote after burying a murdered settler. I read it aloud.

Grimly the troopers stood around
That new-made forest grave,
And to their eyes that fresh heap mound
For vengeance seemed to crave.
And one spoke out in deep stern tones,
And raised his hand on high
For every one of these poor bones,[5]
A Kalkadoon shall die.

'I think he must have been a better policeman than a poet, Blue.' I agreed.

He continued. 'The Kalkadoons challenged him to fight them in the hills. Many whites combined with the native police to join the battle at a place near Daniel's family's homestead called Battle Mountain.'

---

5. Humans have about 500 bones. Some people estimate 500 Kalkadoons died. A prophetic poet!

'Did Ernest or Daniel go, Holas?'

'Not as far as I know, Blue. If they did, they didn't tell me. Daniel was an honorary magistrate for the district, so I expect he knew about it in advance, but he was a little vague when I asked him. I guessed that some of the Kalkadoons who died worked on his family's station or on Quamby so I didn't pursue the matter.

It all seems slightly ridiculous thinking about it now Blue, but Urquhart[6] started proceedings by ordering the Kalkadoons to "stand in the Queen's name" and of course, they did not. They sheltered behind rocks and pelted the whites and police with ant-hill fragments and spears. Urquhart ordered a cavalry charge. The horses couldn't mount the hill so it was useless. Queen Victoria, represented by the colony of Queensland, won eventually because the Kalkadoons left their rocks and charged. They were shot to bits by rifle fire – some said 900 died, but I doubt that. I don't think there were 900 Kalkadoons in the district.'

'Quite different to your experiences further south, Holas?'

'Entirely, Blue. This was a set battle following British military rules and I don't know of it happening in any of the colonies before then. One of the interesting things is that most of the police were blackfellows from other districts of Queensland. As I said, I think Daniel regretted it but he never mentioned it directly.'

'What happened to Daniel, Holas?'

'I didn't seem to have much luck with my partners. Like Fitzherbert Brooke, he died too early – in 1894 – although Daniel was never a financial partner – we paid him a salary and an annual share of the profits. Daniel died at the same time that we were winding up the partnership. He was no longer with us. Frank Brush, a Melbourne investor was my partner then and he was at Cloncurry when Daniel died on his family's station. Jack Phelps had dropped out several years before when Frank bought his share.'

'What did Daniel die of Holas?'

'I never discovered. Frank said it was recorded he died from natural causes after a short illness, so there was nothing suspicious.

---

6. He was rewarded. Urquhart was Police Commissioner in Queensland and Administrator of the Northern Territory.

Frank didn't know until a week after Daniel died. It was a shock. Some said it was a heart seizure, but it was speculation.'

'How did you and Daniel keep in touch with each other Holas?'

'We used wires and letters Blue. Daniel was always elegantly brief. His letters were like wires.

I can still remember one:

Quamby
Via Cloncurry
5th June 1886

Dear Mr Sadleir,
I have dispatched 300 bullocks to join 700 from Mingara at Boulia bound for Marree.
Drover O'Malley will wire you from Boulia.
Yours in good faith,
Daniel McDonald.

Mingara was part of Barkly Downs station in 2011. I telephoned the homestead. The governess answered.

'The manager and just about everyone else is away at a camp draft for the weekend. I can't tell you much about the place historically; I've only been here a month or so. The manager hasn't been here long either, only about four years, but he may be able to give you a few pointers.'

She told me about the place. It was one of a group of stations owned by the Stanbroke Pastoral Company. There were about 25 staff and it ran 75,000 cattle on 1.2 million hectares.[7] There were no outstations. The staff lived at the main homestead and travelled out with vehicles and horses in trailers for mustering and handling at various yards and waters on the run. A helicopter did the first mustering and ground staff with horses and a few motorbikes settled the cattle and walked them to yards.

She was from New South Wales and a trained teacher. She liked it there. 'My special love is the poddy calves. I look after them. The kids are good too.'

A couple of days later I reached the manager by telephone.

---

7. 2,4 million acres or 3,780 square miles.

John confirmed that McDonald's Camp, as I described it, '16 miles south east of Barkly Downs and 22 miles south of Flora Downs', was on Barkly Downs country, but he hadn't noticed anything there to suggest it had once been a homestead.

'On the Mingera Creek?'

'Yair.'

'I know the country you mean, mate. I've flown over it regularly and there's nothing there.'

I was disappointed, but not surprised. We talked about the station and the seasons.

'The last two have been terrific, but we had years of drought before that; we got rid of cattle and we're just building up again. It'll take a year or two to get back up to the 50,000 breeders that I think is about our stable limit.'

'We've got one permanent helicopter musterer on staff. If we need to, we get contractors in to give him a hand or to let him go away for a spell.'

'It's not the cattle that are my worry. They're easy. It's the people.'

'This place is on the Barkly Tablelands, we're just on the Queensland side of the Northern Territory border where most people think the Barkly Tablelands are. This is some of the sweetest country in Australia.'

I thanked him and wished him well.

'You're right mate.' And we hung up.

I told Holas about the helicopters.

'Blue,' he said, some of your stories are too ridiculous for words.'

9

The Adelaide Years

The family moves from Tasmania by ship to Adelaide. It is an exciting and anxious trip. They adapt to new servants, schools, public transport and town life. Four more babies arrive. They have a telephone. Nicholas has help from the police in locating his sister Helena, but he can't find her. He thinks she may be in the Dutch East Indies or in another part of the British Empire.

Great-Grandmother talked about the move from Quamby to South Australia: 'Joseph Phelps was wonderful in his displacement of us. I always thought him to be cold, but when he finally decided that the Quamby homestead would be sold along with all the land, he wrote to me from Ireland telling me that he had set aside £5000 for me to provide suitable accommodation for Nicholas and the family in a suburb of Adelaide of my choosing. He and Nicholas had decided on Adelaide as a place for us to live because of the new Broken Hill to Adelaide railway and the ease of getting from Broken Hill to Albemarle and because Nicholas had decided to sell wool from Albemarle at the new colonial wool auctions in Adelaide.

Before the sale of Quamby, Nicholas and I set off on a steamer to Melbourne and took the regular service to Adelaide. Baby Angela came with us with Maggie Scharpel to help with her care. The rest of the children remained at Quamby, but we planned to be away for less than a fortnight and I knew the household staff would care for the children as if they were their own.

Nicholas and I had agreed we would look for a house at the seaside. I said to him. "Please remember Nicholas, I am from a maritime family – Launceston, Geelong, Plymouth, Greenwich and back to Geelong again. If I cannot have Quamby please can I be near the sea?"

Nicholas said, "Then perhaps it should be the Darling River, my love. There is nearly always water in the Darling at Albemarle."

I think he regretted saying that. And it was a silly thing to say. Jack Phelps was living in the Albemarle homestead.

It was the first time that I been away from my children for more than two days, Robert. And I felt something of a wrench as we left Launceston for the Port of Melbourne, but I'd almost forgotten them completely as our steamer reached Port Adelaide and I gave something of a guilty start when Maggie wondered out loud how the children were back at Quamby. But I forgot about the others soon enough, little Angela had a nasty rash but by the time we returned to Quamby it had cleared up.

We stayed at the Royal Arms Hotel, a short distance from the docks and Nicholas was able to send a wire to the Adelaide offices of Albemarle's agents to tell them we had arrived. The Adelaide office knew we were coming and they knew of our requirements – a large house with sleeping accommodation for nine maturing children, servants and visitors, with pleasant reception areas, functional kitchens, water closets and bathrooms, and a view of the ocean.

Angela and Maggie came with us on the first day in a buggy. Mr Smith, the agent, went with us to the residences he had to offer. He was a considerate man and allowed time to pause at hotels along the way to feed and change our baby. Maggie held Angela when we stopped to inspect residences. She was a great help.

There were few houses for rent, but we saw five in the three days we stayed in Port Adelaide and we chose the one we saw first, 22 Washington Street, Glenelg, a house of seven large rooms. There were separate rooms near the stables and a coach house for a groom and servants. It was tiny compared to Quamby but it was the best that could be had near the sea. Nicholas and I agreed that it would be better to look for a larger house at our leisure when we lived in Adelaide.

We were booked on a steamer travelling directly to Launceston but Nicholas remained in Adelaide and Maggie and I and baby Angela returned to Launceston. He believed his sister Helena may have been in Adelaide and he wanted to look for her. He completed the rental agreement of our first South Australian home and he asked me to list the furniture we needed for the new house via the Electric Telegraph. He and Mr Smith purchased the furniture we needed. There was little of the quality we were used to at Quamby but it was serviceable, Mr Smith was able to recruit a cook, a housekeeper, a housemaid, and a groomsman to manage the house a week before we arrived.'

Great-Grandfather resumed. 'I was doing several things at once Blue. John had written me a letter of introduction to William Peterswald who was the South Australian Commissioner of Police. When I presented the letter Peterswald agreed to see me straight away and came into the foyer to show me to his office. "I'm happy to be of service Mr Sadleir," he said. "Your brother is a fine policeman and a good friend. We exchange thieves from time to time. Is he well? Please pass him my best wishes."

"Well enough I think, Mr Peterswald. I shall see him probably in a day or so on my way back to Tasmania and I shall be pleased to convey your greetings to him."

"I understand from your brother's letter that we are seeking a missing person – a special missing person – your sister."

"Indeed my sister. My twin. She has been in the colonies of Australia for at least 15 years and yet I have not seen her since I left Tipperary for Victoria in 1852. My brother John saw her briefly when she arrived in Victoria. She started a small academy at Ballarat soon after spending some time with him when he was at Kyneton, and then she disappeared. We believed she'd gone to Western Australia – she left an address for a sheep and cattle station in the Roebourne district but our letters to her were returned unopened and John later discovered from a colleague in Western Australia that she had never been seen in the colony. We have no firm evidence, but making some assumptions about most coastal vessels on their way from Melbourne to Perth's port of Fremantle calling at the port of Adelaide, it may be likely that our sister Helena abandoned her quest for work in Western Australia and debarked here."

"How long has your sister been missing, Sadleir?'

"I'm ashamed to say it is more than 16 years since my brothers John and Richard saw her in Melbourne when she told them of her intentions in Western Australia. I believe it was an unhappy parting. John prevailed on her not to go and she grew angry and wilful."

"I see you have a photograph, Mr Sadleir. Is it a good likeness?"

*Helena Sadleir, Nicholas' twin sister,
photographed in Melbourne during the 1870s*

"One must assume that it was when it was taken, Peterswald. I am able to confirm it was her. I noticed that it was taken in Melbourne so I suppose we are looking for someone who looked like this more than 16 years ago."

Peterswald looked sober. "This is no small task Sadleir. Naturally, we are pleased to help. I shall have our police records checked, we have access to most shipping records and I will assign a Constable to make enquiries of our schools – you mentioned your sister's calling in teaching and education – but it may take at least a month for reports of satisfactory investigations to reach me – and, I must warn you – it may tell us nothing."

I rose and handed the Police Commissioner the photograph of Helena. "John and I will be grateful for whatever help you give us. I shall be returning here in a few weeks to a household with my wife and family at Glenelg. Please do not hurry the investigation. For all we know, Helena may have left this continent entirely for other British colonies. If she repatriated to Ireland I think we should have known."

Peterswald walked with me to the door. "I'm happy to have been of service, Sadleir, and I look forward to less formal talks when you settle at Glenelg. Your brother John tells me you enjoy greyhound coursing. I shall be pleased to introduce you to the president of the club and others. We run several courses a year on the plains north of the city. Please call on me again when you are comfortably settled and if you have not heard from me within a month. I have your new address at Glenelg and your card indicates your current abode in Tasmania. If I find anything urgent immediately, I will wire you there. In any event I look forward to welcoming you here soon."

And so I returned to routines, Blue: the ordering of furniture and the accounting of wool coming by rail from Morgan on the River Murray to the newly established wool stores at the port of Adelaide to be ready for sale at the new colonial wool auctions; and a trial shipment of sheep by rail from Broken Hill to a sheep auction in Burra Burra in the colony of South Australia. We had never used rail for sheep before from Albemarle – Queensland cattle from Marree to Adelaide, yes, but never sheep. We were wary about the expense, but Jack Phelps was keen to try new ideas.

I returned to Melbourne by train and re-joined Madam and the children at Quamby to help with our final departure. I was looking forward to being closer to some of our markets, so I wasn't entirely regretful at leaving our beautiful Tasmanian home.'

I talked to Great-Grandmother about the move and the children.

'And the journey to Adelaide from Launceston with the family, Great-Grandmother?'

'An exciting adventure and a nightmare in equal measure for the children, Robert. Elsie and Jim sulked for weeks when we told them of the new home they were going to. They had strong friendships and they seemed to rehearse their mannered way of looking sadly tragic at leaving their nearest and dearest. Elsie was especially good at it. Mary ignored them and took charge of the smaller children. She recruited Georgie and Nicholas to form the remaining children into port and starboard watches and briefed them on their responsibility of watching each other to prevent them going overboard. She drew lines in chalk on the veranda at Quamby and exhorted Georgie and Nicholas to practise controlling their bands at all times – it was delightful – remember the five children they supervised were five years old or younger. In the event, three servants from Quamby came with us for the journey to help with the children, but Nicholas and I always gave Mary credit for the discipline she imposed on her little brothers and sisters to the extent that we lost no one at sea. Nicholas was seasick. He was of no use at all.

It seems strange, but the saddest part of the move for me was saying goodbye to those three women as they left on the steamer to take them back to Launceston. They'd been valued friends as well as servants for 13 years and I knew I would probably never see them again.

I hated leaving Quamby at first, Robert, but I gradually came to appraise the advantages of living in a city. Glenelg was about eight miles from the city of Adelaide and there was a regular railway service. Elsie, Mary and Jim could go up to Adelaide to school on the train as soon as we arrived from Quamby, and eventually all the children travelled on the train to school. The boys went to the Glenelg Grammar School and then St Peter's College (except Toler who won a scholarship to Prince Alfred College) and the girls went to Miss Gilbert's school at Glenelg for their primary schooling and to the Advanced School for Girls in the city later.

Servants were easy to find and not all of them had to live in the house which made more space for us and the children – the cook and several of the housemaids travelled to work at our house each day.

A nanny lived in and the driver and groom had his special quarters beside the stables.

And of course there was the sea. We all loved going to the beach and in summer, the children could swim after school and they sailed in small dinghies. So, after about a year, the charm of Quamby faded.

We had four more babies at Glenelg: Aubrey Toler (we always called him Toler – a clever boy), Alicia Grace (called Tisha or Tish – a talented painter), John Raimond (Jack) and Kathleen Ross (Kath) – our two babies.

Times were good. Nicholas travelled to Albemarle regularly on the train to Broken Hill and via coach to Menindee, and he went back to Tasmania in the first years to finalise the sale of the farms at Quamby, and he was able to avoid seasickness for some of the journey at least – he caught the express train from Adelaide to Melbourne. He didn't go to the runs in North Queensland often but occasionally he took the train to Marree and met drives of cattle coming down. There were railways everywhere.'

Nicholas resumed. 'And Madam forgets, besides telegraphic wires we had the telephone. At first, one had to go to the Post and Telegraph office to make a telephone call, but later we had the telephone installed at Glenelg and at Albemarle (I let a contract to run the wires and poles out from Menindee and on to some of the outstations – Wheelers Well and Victoria Lake – it was about 100 miles). It was a wonderful convenience because one could dictate wires over the telephone to the Telegraph office. From Glenelg I was able to speak to our Adelaide agents any time during office hours to explain things if I had a complicated instruction, or I could send a wire for simple messages – or to confirm a telephoned instruction. Madam decried the fact that my delivery on the telephone was not sufficiently theatrical. She suggested elocution lessons.

Intercolonial telephone calls were expensive so I used wires to keep in touch with Jack Phelps at Albemarle, Frank Brush in Melbourne and Daniel and Ernest at Cloncurry and our agents in Melbourne and Launceston. On a couple of occasions Daniel telephoned me from Cloncurry, and that was something of a thrill for us both, but we hardly ever did it. It cost too much.

Peterswald, the Police Commissioner for South Australia took me coursing as he had promised. We drove together in his buggy to a run called Buckland Park for a good day of coursing, but on the way there he told me what he discovered about Helena. "Sadleir," he said, "you are right to surmise that perhaps your sister left a ship in Adelaide en route to Fremantle. The Department of Education has a record of a Helena Sadleir being appointed as teacher in charge of a school at Macclesfield in the Hills to the south-east of Adelaide but she never took up her appointment. That was in March 1874. There was a note on that record that she had been unable to take up her post because the Governor of South Australia had persuaded her to establish a school at Pine Creek, a gold-mining settlement in the Northern Territory. It was apparently a special project of his. The records of the school indicate that she was there for less than six months and there was a note in the records in Adelaide that *she had resigned and moved to Palmerston*. And there the record stops, except to say that there is a rumour, but no shipping record, of her embarking on a Dutch mail packet to Batavia, but I'm sorry to say there is no record of her having debarked in Batavia, and further, my source has doubts about the rumour."

"Lost at sea?"

"Perhaps. But shipping records in Dutch colonies were not always reliable, so she may have left the ship in Batavia or she may have gone on to Rangoon, its next port of call – and before you ask, there was no record of her debarkation in Rangoon."

"So she could be anywhere. She may be still in one of our colonies."

"Yes."

"Have you any theories, Peterswald?"

"She sounds an adventurous and mature woman Sadleir, and I remember seeing the Governor's notes of his good impressions of her capacity to manage a school when he asked her to go to Pine Creek – he regretted his inability to employ her as a tutor for his own children at Government House in Adelaide. It is just likely that Dutch colonists persuaded her to teach in Batavia. Several other teachers left from the colony of Queensland to do that. The Dutch settlers were desperate to establish educational facilities as an affair of their own. There was less help from the Dutch Crown in the way that our

Queen endows schools here. But I really am speculating. However, I have followed up that line of enquiry. An old colleague is serving in the Batavia police and I have written to him to see if he can discover anything. So there may or may not be more news."

"Could she have been lost at sea without record?"

"Oh yes. It happens frequently. But wait and see, Sadleir. We may get more from Batavia."

Soon after we settled in Adelaide, I got news from John in Melbourne that our mother had died at Brookville House. John and I had not seen her for 37 years and Marshal had been gone from Ireland for 18. She had more than a score of Australian-born grandchildren she had never seen, a son, Richard, buried in Melbourne, and a daughter who had been missing for years and who could easily be dead too. She was 90 when she died. I was 54 and so was Helena if she was still living.

For the first time, some of the children were able to come with me to Albemarle during their holidays from school in summer. Madam remained at Glenelg with the younger children who jealously resented the departure of their older siblings and begged to be allowed to come. Georgie, your Grandmother, Blue, was most plaintive in her protests. "Pater, I am nine," she said. "You told me I was the best rider of all the children at Quamby. Mary is a useless rider. How can she help you with mustering? James will be of no use at all. All he will want to do is shoot rabbits and that is cruel. Pater you need me as a reliable and experienced horseman to help you and Uncle Jack at Albemarle."

"Next year, Georgie." I said. "When you are ten." And when Georgie finally came to Albemarle she loved it, and she was as useful as she predicted. I remember telling her how sensible and tireless she was at working sheep in the yards when we were classing the ewe hoggets and she said, "Yes Pater, I think Uncle Jack is getting a bit old for this. Perhaps I can leave school and take over from him as manager?"

'Well she married a farmer, Holas.'

'Did she? Well that was good. Where did she go?"

'To Tintinara in the ninety-mile desert south of Murray Bridge in South Australia.'

'Terrible country, Blue.'

'Yes but it had promise with new fertilisers when she went there. She died from there when she was 32. She was the only one of your daughters to marry.'

'That is sad. She was a lovely girl. But they all were. And Mary was by no means useless as Georgie suggested. When the children first came up with me we used to stay in the homestead with Jack Phelps and Mary became assistant housekeeper. She used to pick flowers from the garden every day and arrange them freshly in vases in the homestead, help the cooks make jam and chutney from the fruit she picked, and when she finished that she used to help the bookkeeper tidy up the accounts of the station store. Mary wasn't really an outdoor type but she loved being useful. The household staff loved her. They could loaf while she did their work.'

'And what about Elsie and Jim?'

'Jim loved shooting and he was good at it. He had a pea rifle and I supplied him with as much ammunition as he could use from the station store because the rabbits were starting to overrun the place since they had arrived in the district 10 years before. We had men on poison carts as well, but Jack Phelps and I took the view that any rabbits eliminated would help preserve the sheep feed we needed and that Jim was making a contribution. There was no need to slaughter old rams for dogs' meat when Jim was about. The station dogs had more rabbits than they could eat. Jim was good with wool too. He learnt a lot from the sheep classer in the yards at the woolshed.

'Elsie spent most of her time on Albemarle with horses and dogs. She used to get sentimental about them and she would swim with horses in the river on hot days. She made up small poems about the dogs telling us of their heroism in dragging Aboriginal babies from drowning in the Darling River or of horses galloping from Menindee with urgent news of sheep sale prices at the Flemington markets in Melbourne. Elsie especially loved riverboats. On the second holiday Sam Curtis took her to Morgan with wool and back.'

'She went by herself?'

'Yes. She was about 16, and Sam protected her like a hawk. Nevertheless, Elsie and I agreed that the adventure was best kept from Madam. Elsie returned with a much expanded vocabulary.'

'Did you always go with the children on the train and the stagecoach, Holas?''

'In the beginning I did, Blue. But usually Elsie or Jim or Mary would be with several of the younger children and they were able to take charge because they knew what to do and the railways people and the coach people knew them and would make sure they got to Menindee. We usually took the night train from Adelaide, the coach would leave for Menindee the following morning and there would be a buggy to meet the children in town with a three-hour trip out to Albemarle. So it took about two days. But that's when things were running smoothly. On a few occasions, we were held up because the line was washed out after good rains and sometimes there would be derailments, but by and large, it was a good reliable service from Adelaide.

And there was always the river as a highway when it had water, excellent for freight, agreeable for elderly tourists and much too slow for bouncing children. Georgie, Robert, Nicholas and de Vere took the train once to Morgan and caught one of Sam Curtis's steamers on the way up to Albemarle to get wool. I think the novelty lasted about a day and then they got bored. It took more than a week to get to Albemarle. They were overjoyed to see the Albemarle woolshed – and so was Sam Curtis – he tried to entertain them with card games. Sam was never any good at cards, he was unused to children, they ate nearly all his rations and they cleaned him out at cards.' He chuckled.

'We had huge floods in 1890. The homestead was safe but we had to move some people and a lot of livestock to higher ground. I can remember some of the paddle steamers left the river and went to their destinations directly. They marked the way with poles driven into the mud carrying flags so that they could find their way back to the Murray to turn up or down river to Morgan or Echuca We had good sheep feed for two years afterwards.'

'And how were the markets, Holas?'

'Wool prices had been decreasing slowly for about six years. Albemarle wool averaged about £25,000, reducing by about 1% each year. I was able to reduce costs of selling by supporting the colonial wool auctions and withdrawing from our traditional London market. The cost of freight to London for two to three thousand bales of wool every year from Albemarle alone was no mean thing. We sold the Albemarle and the Boondarra wool in Adelaide, floating it down the rivers to Morgan and then via rail to the wool stores at Port Adelaide. The net price compared favourably with London – although both markets were falling, but wool buyers could ship wool directly from Australia to their mills. And the woollen mills of the world were no longer just in Yorkshire. America, Continental European countries and Japan were all building mills and opening up new markets for cloth. So we were able to reduce the expense of freight to markets Blue, but overall the last decade of the nineteenth century was not a happy time for me or for stockowners generally. Prices kept falling, we had a depression, a general strike, a shearers' strike (that nearly became a rebellion), a bank crash and the start of a terrible drought.'

10

Ruin and Drought

A depression emerges, Nicholas loses his equity in the Queensland cattle stations, he spends more time at Albemarle and a son kills himself because he is in trouble at school. Anna starts a clothing wholesale business. The children visit Albemarle for school holidays and enjoy themselves. The family moves to a smaller seaside house. Drought strikes Albemarle. Livestock diminish by 90%. Enlistment in the Boer War, movement to other occupations, and the ruinous results on the Barkindji people assists Albemarle's depopulation. Nicholas compares it with the potato famine in Ireland. He describes the erosion, devastation and dust storms.

Nicholas said, 'One of the first and worst things about the 90s was that Joe Phelps died in Ireland. It was sad for me. I was closer to his brother John in the early years but Joseph probably taught me more about the organisation of people and that helped during the troubles we had with shearer strikes and the large numbers of unemployed people in the depression. He was 73 when he died. I wrote to him regularly with reports of station affairs until his death.

I never met Joseph's widow. She remained in Ireland. I sent this brief letter of condolence to her home at Willowbank, Limerick, Ireland.

22 Washington Street,
Glenelg
South Australia
5th May, 1890

Dear Mrs Phelps,
Firstly, allow me to pass on my personal sympathy and the condolences of the many people at Albemarle to you on the death of your husband Joseph. His absence is much lamented in the Western

districts of New South Wales where he was highly respected as a champion of the region in the New South Wales Legislative Assembly. I personally have lost a fine master and friend. Mrs Sadleir asks me to pass you her deepest sympathy. Joseph was kind to her in her early days at Albemarle, was a gracious guest at Quamby and was kind and considerate to her and the children when we left Quamby for life in South Australia.

Please feel assured of my continuing loyalty to your well-being and the sober management of the affairs of Albemarle. I await your instructions.

Yours sincerely,
Nicholas Sadleir.

Mrs Phelps acknowledged the letter and told me that she had placed the administration of Joseph's assets with her relative A.C. Ferguson. He and I corresponded regularly. We advertised Albemarle for sale by auction in Melbourne in 1893 and later withdrew it. We were in a depression. Prices were low. It was not a good time to sell.

Ferguson came to Albemarle in 1894 to appraise a proposal from Jack Phelps to install mechanical shearing in the woolshed. He supported Jack's idea and we put in a mechanical plant driven by a steam engine. He came to the West Darling in an excellent season, the country looked wonderful, there had been a slight increase in the price of wool, and so he was not visibly conscious of the strife of depression or of the strikes. He was a nice chap but pretty green. He knew nothing of Australian pastoralism. Fortunately, he left the management of Albemarle to me. Soon after his visit Mrs Phelps died leaving a 15-year-old son. We continued to trade as the Estate of J. J. Phelps. Jack became one of the executors while he was working with me at Albemarle. His father Robert kept an eye on the estate as well from Valverde.'

'Tell me about the strikes, Holas.'

'Well I knew more about the shearers' strikes, Blue. But in many ways the strikes were all part of a series of political and economic calamities. It may be fair to say that English bankers started it because of a depression that started in the Argentine – all of England got a cold and prices for just about everything dropped, and that included goods from Australia – especially wool. Ship owners in Australia and New Zealand started to feel the pinch, they got anxious about unionism and so they declined to pay a demand for increased wages from ship's officers. Then the officers refused to sail. They went on strike, and to add to it, general seaman and waterside workers went on strike as well. The strike even spread to the New South Wales coal mines.'

'Well what happened, Holas?'

'They lost, Blue. Lots of good strong men were unemployed and they moved in to take the work of the people who were on strike. The striking workers had families to feed and so eventually they had to go back to work, and when they did, employers offered them less pay, sometimes 30% less.

It was a terrible time, and then sheep owners in Queensland offered their shearers 30% less per hundred and the shearers went on strike. And it wasn't just the money, Blue, the shearers formed a union and they refused to work beside shearers who were not members. This strike was more vexatious. There was trouble in most of central Queensland. Shearers armed themselves, they menaced non-union labour, they burned woolsheds and they lived together in large camps. The police and the army were everywhere.

'How were things on the Darling then, Holas?'

'We had strife at Albemarle. Our shearers refused to start with the agreement we had in place and they camped downriver, hoping that we would relent and pay them what they asked. I was worried for a time and asked for police protection. But it all came out all right in the end. We renegotiated and the shearing got done.

It was fun in retrospect, Blue. We had shearers' union members chasing free labourers on paddle steamers up and down the river. I even sent this telegram to the Pastoralists' Union of New South Wales: ***Seventy unionists chasing free labourers in steamer, and holding this station by force.***

Everyone was expecting huge fights, but it came to nothing.'

'Some people say that the camp at Barcaldine in Queensland was where the Australian Labor Party started, Holas.'

'Yes. I heard that too Blue. But it didn't do too much good at the time. The striking shearers starved in Queensland, most sheep were shorn with non-union labour and the ringleaders of the strike went to prison – some of them for more than three years.

It was miserable here on the Darling even forgetting the strikes. Prices were low and there was no work. There were lots of men travelling for work. We weren't looking for extra workers, if anything it was the opposite, we were having trouble paying the people we employed because of falling wool and livestock prices, so in the end the travellers didn't bother to beg for work, they just wanted food. And we had a particular problem on the Darling, Blue, because we were close to Broken Hill. People were put off at the mines and there was a strike there as well, so people moved up and down the river looking for whatever they could get to eat. We had queues of people at our cookhouses every day. We had to feed these poor devils, it was an additional expense, but we had to do it, and then we got another shearers' strike in New South Wales – three years after things started in Queensland. Have you heard stories about the burning of the paddle steamer *Rodney*, Blue?

'A little. I don't have much detail.'

'It happened on the Darling a bit south of Albemarle. The *Rodney* was carrying strike-breaking shearers from Melbourne. She had picked up the men at Echuca to take them to Tolano, one of our southern neighbours, and a camp of shearers ambushed her just out of Pooncarie. The striking shearers captured the boat, threw the free labourers off (the strikers called them scabs), soaked bags of chaff in the hold with kerosene and set the *Rodney* alight. No one died but it was a terrible waste of a good boat. It was one of the fastest things on the river. There were prosecutions for riot and a few people went to prison, but I think the unionists who burnt the boat were acquitted in their trial at Broken Hill. A lot of the ordinary working people were on the side of the unionists. Reasoned argument prevailed. I was even quoted in *The Worker* in *Brisbane* by W. G Spence, a unionist.

I have also a letter from the manager of Albemarle, on the Darling, who says that he had 40 machines and 28 hand shears at work. He states that the machine shearers received in cash thirteen-sixteenths of their earnings and the hand shearers twelve-sixteenths. It must be remembered, however, that he

found all the tools, combs and cutters for the men shearing with the machines, whereas the hand shearers had to find their own shears, and other requisites. As the men are now charged for most of the combs and cutters it will be seen that the case mentioned only proves that the men earn no more with the machines than with hand shears, whilst the profit of the squatter is very much higher! I have quoted these statements because they are, no doubt, the strongest the other side have, seeing that they were written with the intent that they should be used as an argument for charging men for combs and cutters. It is quite clear that the P.U. had other reasons than those of a profit to be made by a reduction in machine sheds, and that their object is to kill the bushmen's organisation if they can, and then do as they like. They will not succeed in their aim, as old shearers remember too well the tender mercies of the majority of the rank boundary riders who pose as station owners.

'I don't quite follow the reasoning of your calculations and his rebuttal of them, Holas.'

'Well in a way we were both right, Blue. I mentioned it to you to demonstrate that we were having a reasoned argument, and not running around with police to catch shearers burning wool sheds or paddle steamers. My point was that we paid the machine shearers more per head because they took off about half a pound of wool more than the hand shearers. He countered by saying that the pastoralists' union was making a case for shearers to provide their own combs and cutters and so the shearers using the machines would be no better off. And there were many other arguments. Some of my pastoralist colleagues thought the rate per head should be reduced because machine shearers could do more sheep in a day.'

'Well what was the result of the strike?'

'Many of the free labourers were not good shearers and so we, the station managers, and the New South Wales Pastoralists Association, decided to meet the union at least part of the way with a conference agreement. One could say that the striking shearers were partly successful. We succeeded too. At Albemarle, after a short strike, we offered to pay the shearers according to the agreement and we had no trouble from the union or the Pastoralists Association.

We shore 115,000 sheep and sent away 2000 bales of wool. I had no interest in other stations at that stage. I had ended the partnership at Boondarra four years before and Frank Brush, my old Melbourne partner, carried on with Quamby and Mingara in Queensland, so Albemarle was really the only station influenced by the shearers' strike I managed. Jack Phelps and I agreed on employing good shearers and we improved the wool shed and the quarters.'

'Why did you give up your interests in Quamby and Mingara, Holas?'

'I didn't want to, Blue. It was the last thing I wanted to happen. They were both new stations. Neither had reached their potential, we had more waters to put in and we had plans for holding paddocks on both the runs but the bank crashes changed everything. My cash reserves disappeared in deposits I held with minor banks that had closed their doors and the bank we used for the firm *Sadleir and Brush* refused us credit. We were not making profits. We were still investing – putting money from cattle sales back into improving the stations. But the value of cattle halved and so we were left to the resources of Frank Brush who had longer pockets than I did.

Do you remember Madam telling you about the £5000 Joseph Phelps set aside for our housing in Adelaide when we left Quamby?'

'Yes.'

'Well I took that to contribute to developments on Mingara. We spent it on contractors to sink bores and wells, equip windmills and to put in tanks and troughs. Frank and I agreed that the £5000 would rest as my increased shareholding in the firm overall. We paid wages and expenses from a joint account with our bank at Cloncurry.'

'What did Great-Grandmother think of that? I understood from what she told me that the money was for her to spend on housing.'

'Well yes. That is likely to be her version, but we did talk about it at the time and I was able to persuade her that the money was best used in a safe investment at Mingara and that we would get it back with improved returns within two years.'

'And five years later it seemed that you had broken your promise, Holas?'

'Yes. But I was able to use it as a debating point with Frank. I took the train to Melbourne when Ernest my nephew sent me a wire from Cloncurry telling me the bank had dishonoured cheques paid to station hands at Quamby and Mingara. He wired: ***Wage cheques dishonoured. Please advise***. He must have been mortified. It was a terrible thing not to pay wages, and often station hands didn't collect against their balances on the station books for more than a year. There was no need for money on the station. Tobacco came from the station store and was debited against their balances. They saved the remaining cash for holidays or new lives. It was a lot for them to lose.

Frank knew what I wanted when I walked in to his offices at St Kilda. He had seen Ernest's wire and I had wired ahead seeking a meeting. I had travelled on the overnight express to Melbourne from Adelaide.

"Greetings, Nicholas," he said. "I wish we were meeting in happier circumstances. Our Queensland cattle, which number nearly 15,000, are worth less than half their value two years ago. We are not in drought. Ernest reports a sufficiency of feed and water and our cattle are fat, but if we set out with a drive of 1000 fats to Marree and then by rail to the Adelaide market I doubt the revenue from the sale would cover droving and rail freight expenses."

"I agree Frank." I said. "My funds are exhausted. The £3000 I held in bank deposits went when my banks closed their doors. How have you fared?"

"Perhaps better, Nicholas. The value of my holdings in real estate in Melbourne has halved, but I still have some cash (not in a bank but in the safe behind me) and the Union Bank still gives me a reduced line of credit against my lands in Melbourne."

"Can you cover the dishonoured cheques in Cloncurry, Frank?"

"I can, Nicholas."

"In what circumstances will you do it, Frank?"

"If I continue to conduct Mingara and Quamby on my own account."

"Are you proposing to buy my share, Frank?"

"Yes. Although I must point out that the value of the livestock and the improvements on the runs is much diminished."

"I agree our cattle and sheep are worth less. But we do not have an equality of equity. I believe our accounts will show that the value of the cattle I contributed exceeded the value of yours by £5000 when we formed the partnership and that I contributed a further £5000 for contract work on improving waters at Mingara."

"I have made some calculations, Nicholas. I believe that the value of our investments in livestock and improvements to our stations has diminished considerably. Thus your advantage in equity in the firm over mine is much less too."

"Frank, I invested $5000 on bores at Mingara that were set aside for the purchase of a house in Adelaide. We continue to live in a rented house and I need at least to recover that to invest in more suitable and larger accommodation. I now have 12 children and another is on the way. If we agree that the value of our investments in livestock and improvements at Mingara and Quamby have halved, then I will accept £5000 in exchange for my share of the livestock, plant and improvements at the two stations. I admire your courage and determination in continuing with these two runs and I wish we could maintain the partnership, but I have no more capital to risk, and a family to keep. I will have to be content to do that with the salary I have from the estate of Joe Phelps for the stewardship of Albemarle."

Frank smiled. Then he reached across his desk and took my hand. "I agree, Nicholas. I shall miss your judgement, friendship and fair dealing. I will send a wire to Ernest at Cloncurry immediately assuring him that the cheques will be honoured, and offering him a position with me as sole proprietor. May I commission John Hamilton, our solicitor, in drawing up the necessary papers? I will bear the expense of that."

And so we went to lunch. Our meeting to dissolve the partnership took less than half an hour. Frank and I got on with things. We always had. It was only on the train on the way back to Adelaide that I had some sadness. I had invested, managed and planned in the Barkly region with Mingara for 14 years and at Quamby near Cloncurry for 12 with Daniel McDonald and Jack Phelps and then later with my nephew Ernest and Frank Brush. I knew I would miss the country and the drives of cattle coming down to Adelaide via Marree.

Frank and I made this newspaper advertisement in 'The Queenslander', Brisbane on Saturday, 3 March 1994:

NOTICE is hereby given that the PARTNERSHIP

heretofore subsisting between us the undersigned, NICHOLAS CLARKE SADLEIR, formerly of Quamby, Launceston, in the Colony of Tasmania, now of Menindee in the Colony of New South Wales, Gentleman, and FRANCIS SAMUEL BRUSH, of St. Kilda, in the Colony of Victoria, Gentleman, in the Business of Sheep and Cattle Farmers and Graziers, at the Stations or Runs known as Quamby, Timboroo, and Mingara, in the Burke District, in the Colony of Queensland, under the style or firm of ' Sadleir and Brush,' has this day been determined and DISSOLVED by mutual consent. All Claims against the said late Partnership will be paid and discharged by the said Francis Samuel Brush, who will receive payment of all Debts owing to such Partnership. Dated this Sixteenth day of February, 1894. NICHOLAS C. SADLEIR. Witness to the signature of Nicholas Clarke Sadleir— John F. Hamilton, Solicitor, Melbourne. F. S. BRUSH. Witness*to the signature of Francis Samuel Brush— John F. Hamilton.'

'Why did the advertisement list your address as Menindee in New South Wales instead of Glenelg in South Australia, Holas?'

'That was Madam's idea, Blue. Not to put too fine a point on it, she distrusted my financial acumen because I had gone bust in Queensland. And so when I returned the £5000 to her for the care of the children and the purchase of a household, she invested it in her own name, used some of it to start a business and to finance a loan with a bank for a large house at Robert Street, New Glenelg.'

Great-Grandmother interrupted. 'You exaggerate, Nicholas. Some of the money went to the account you had to pay school fees for our children, and I used some to pay off debt. I was not as trustful as you were of Frank Bush and I feared that creditors against the old affair you administered in Queensland, Nicholas, could seize our new home in Robert Street. Mr Price, our bank manager, advised it. It was extremely unusual for the times – for the wife of the family to be named on mortgage documents – but it went ahead without fuss. The parliament of South Australia had just passed an Act to enable married women to own property. My suggesting that your address be Menindee in the advertisement was simply a way of protecting our new home with additional precautions. It was a lovely new house of

15 rooms – not as good as Quamby – nothing could be, but a fine house.'

'Unnecessary, my dear. But I did not oppose it.'

'That is as may be.'

'And Kathleen your last born arrived in the new house in Robert Street?'

'Yes, with great delight, Robert.'

'Kathleen seems an unusual name for your family. It is not like the other girls' names in your brood – Eliza, Mary, Georgina, Angela, Alicia – they all seem to appear in family pedigrees from the Sturgess, Hunt, Clarke or Sadleir lines. But this was the first Kathleen. It is really Irish. The other girls had English names. Why was it so?'

'Your Great-Grandfather named her, Robert. I remember asking him the same question and I got an unsatisfactory answer. But Kathleen was a lovely girl – our baby.'

Nicholas remained silent.

I said, 'I think I can imagine why. Was it something to do with your sea voyage to Australia Holas? Or perhaps a special someone from Tipperary?'

They were mute.

I changed the subject. 'You mentioned starting a business, Great-Grandmother. Will you tell me about it?'

'Willingly, Robert. But it was no great thing. I called it the Cellular Clothing Depot. It was in the centre of the city of Adelaide. I sold bolts of cloth called *Aertex* to retailers in Adelaide and in country towns in South Australia, and I sent consignments to Broken Hill regularly. It didn't start it until 1897. Kathleen was weaned and I could leave her at home with Tisha and Jack in the care of the household, and the rest of the children were at school.

'I've worn *Aertex*, Great Grandmother. It was a light cotton material with holes in it and we used it for tennis shirts, but why was it called cellular clothing?'

'Oh that was the original name, Robert. The Cellular Clothing Co started in 1889 in Lancashire. The proprietor of a cotton mill had an aunt who believed in the health-giving qualities of circulating air in clothing and so he consulted physicians and started to experiment by weaving air spaces into cotton fabric and promoting its healthy properties. He did it wonderfully. *Aertex* went all over the world in bolts of cloth. Fortunately for me, there was no distributor in Adelaide. I'd noticed the cloth on a trip to Melbourne so I wrote to the company in England and they agreed to supply me with cloth at wholesale prices. As I said, it was not a grand affair but it helped with school fees and household expenses – we were unused to them, everything went on the station accounts at Quamby – and it supplemented Nicholas' salary from Albemarle. I ran it for 12 years. Women could thrive independently in South Australia. We were able to vote in 1895 and stand for parliament – the first colony in Australia.'

Nicholas continued with a lowered voice. 'If we talk softly, Blue, I doubt Madam can hear. This was the worst thing of the 1890s. My namesake, the lovely boy Nicholas, killed himself with a shotgun in the chaff shed at home at Glenelg. He was 15. It was terrible for Madam. I was at Broken Hill and I got the news by wire. I came straight down on the train to comfort her and the rest of the family, but I wasn't of much use. I needed comfort too. '

Can you imagine why he did it, Holas?

'Not really. It is so hard to get inside a boy's mind. We know he was in trouble at school, and that was the agreed cause of Nicholas killing himself, but to answer your question, I can't imagine anything that could be so bad to cause that. Madam took a different view. When she discovered that he had been falsely accused of misbehaviour at school she wrote to both the newspapers in Adelaide to clear his name.'

I found the article:

The Register, Saturday 14 November 1896
A SAD CASE.
TO THE EDITOR.
Sir — Will you please give the following room in your columns: – It concerns the circumstances surrounding the death of the late Nicholas Sturgess Sadleir on whom a coroner's inquest was held on the 29th of August. Had I possessed the information at the time I would have given it, but did not hear anything which I could call definite until last Saturday, November 6. I am trustee and guardian for my children, and think it my duty to clear my

son's fair name in any possible way. On the evening of the 28th of August I went to St. Peter's College and saw the headmaster, who told me that an undermaster had reported my son and that his reply was "Sadleir is dead." He told me that nearly three years ago he had caned my son for misconduct, and had then told him he would send him away if such occurred again. He did not tell me what the bad conduct was, but I heard of it lately, and found two who were present and gave me a truthful account of the matter. Their testimony was that my son was innocent, and that he protested his innocence at the time, but was not listened to. It seems to me the idea on the boy's mind when again reported was that he would lose his position, and, in fact that life would be unendurable to him. This seemed so terrible that he resolved on taking his life, hoping that God would forgive him. According to the evidence at the inquest the impression left on anyone's mind who did not know the boy might be that he had done wrong, and feared the consequences; in fact, his father at Broken Hill was told that it was something disgraceful. I write this, hoping that, as he was innocent on the first occasion, as proved so to my satisfaction, so he was on the second, and that any of those who formed an opinion on the results of the inquest may possibly modify it if it is adverse to an honourable boy. I do not wish to cast any reflections on anyone, and have no doubt the headmaster did as he thought best and wisest in caning him. We are all liable to mistakes, and sometimes these are fatal. I gave my son into the care of the Rev. P. E. Raynor soon after he came to Adelaide, believing St. Peter's to be as nearly on the lines of a church school or an English public school as possible. One of the masters at the time remembered my son as a dear good boy, and his parents always found him so. – I am, &c., ANNA G. SADLEIR Glenelg, November 13, 1896.

'It must have been terrible for her, Holas.'

'It was Blue. On the anniversary of his death she had a memorial notice placed in the Melbourne Argus:

The Argus (Melbourne, Vic.: Saturday 28 August 1897

SADLEIR – In memory of Nicholas Sturgess Sadleir, third son of Nicholas and Georgina Sadleir, and great grandson of

Henry Davis Hunt, of Cappagh-house, Co. Tipperary, descended from John (de Vere), 16th Earl of Oxford, died 1550, died at Glenelg, South Australia, 28th August, 1896.

Vero nihil verius

"The trustiest, lovingest and the gentlest boy,"– Beaumont and Fletcher

"Misunderstood. And I smiled to think God's greatness flowed around our incompleteness." – E. B. Browning.

"Love is all, and death is nought. We fall to rise, are baffled to fight better – sleep to wake" – Robert Browning.

"Whom the gods love die young."

"A Vere of the Fighting Veres" – Ludlow Papers," by T. Carlyle, Oliver Cromwell's Letters.

"I am the resurrection and the life."

'Extravagant prose, Holas?'

'It seemed to be, Blue. Your Great-Grandmother started her cloth business after that. Perhaps she needed it for sanity.'

He paused and started to talk about Albemarle in the late 1890s.

'I spent more time at Albemarle and less time in Adelaide because Jack Phelps wanted my help. He sought time to develop a retail affair at Corryong in the beautiful country at the headwaters of the Murray River more than 500 miles away. It was near his boyhood home at Valverde near Albury.

For much of the 1890s we had to cope with homesteaders who would pick out 1000 acres or so along the river and start a small block. It was legal. The government encouraged it. They were a nuisance and at times over the years we bought them out, only to have to do it all over again. They cut wood for the steamers, ran goats and a few sheep and cattle. Some of them started pubs and stores and these were popular with the men but not with me – except perhaps for the humorous relief. The Roseborough Hotel was on Albemarle and I can remember three scallywags who got into trouble after I had sacked one of them for drunkenness. It was shearing time. He was a cook for the rouseabouts, and when he had ceased profitable employment with me he fell in with two others who called themselves the Bulldog Push because they used to go about pubs with a huge bulldog. The police finally caught up with them at a place called the Harp of Erin about 53 miles downriver where the rouseabouts' cook had accused the other two of robbing him of his watch and chain and about five pounds. The jury acquitted the thieves partly because the two accused had suggested they were in trouble because the "policeman who arrested them didn't like bulldogs" and they offered members of the jury the opportunity to gambol on the ground with the dog to find out how gentle he was. And what was even more amusing, was the so-called victim said he was Finnish. He called himself David Carter. I think he was born in Ballarat!

'Lucky for them you weren't on the bench Holas.'

'Yes, I suppose so, Blue. I never did much court work. Jack Phelps did most of it. He even sat on hearings at Wilcannia.'

More seriously, there was always agitation to put weirs along the Darling up river and I had to fight it. One had gone in at Bourke. I thought it was a foolish thing. I gave evidence before a Parliamentary inquiry and I can remember writing this letter to Sir Joseph Abbott who was an influential Member of Parliament.

Albemarle,
Menindee
14th Aug, 1899.

Dear Sir Joseph,
Thanks very much for your prompt action on receipt of my wire about the Bourke weir, and for your letter, which I received by last post. It is a most serious thing to interfere or stop in any way these small floods in the Darling, thereby endangering the possibility of navigation by steamers. The erection of two other weirs between Wilcannia and Bourke and the Bourke weir would have entirely impounded all the water that has come down the Darling for the last three years, and made navigation impossible between Wilcannia and the Murray River, depriving all the people who live on its banks, and for hundreds of miles back from the river, of cheap and possible carriage for their supplies and produce going away. The severe droughts we have experienced for the last four years have made carriage by teams almost impossible (camels have frequently been employed to carry on steamer loads when they get stuck in the river).

This country cannot be held and worked to pay its way without cheap carriage, which only is possible by navigable river.

No one can show the least advantage derived to now by the Bourke weir, whereas any number of people complain of the serious disadvantage it is proving to them when floods are low and weak. Weirs may, in time, after hundreds of thousands of pounds have been spent on them, prove of some small advantage to navigation, and only when the entire system is completed. I think the weirs are being put up at the wrong end of the river. If begun where the Murray water when in flood ceases to influence the Darling, and continued at proper stages following up the river, an immediate advantage would be derived, and no one would have any cause of complaint. From my experience of stopping floodwater in the country composed of sedimentary deposits, such floods as 1864, 1870, 1886, or 1890, will, within two days after it reached the weir, cut across some bend or low flat, and will scour out a new channel as deep as the main river channel. Many such new channels have been made within my experience. The Darling channel is not big enough to carry any such floods as I have above mentioned, and when the channel is lessened by a weir, the scour will be greater. I expect to see the river above the weir silt up when those new channels are scoured out, and they will be looked on as a folly. Such can be seen about here, where big creeks have been dammed to stop floods, and new channels as deep as the main creek have scoured out since.

I remain, yours faithfully,
NICHOLAS SADLEIR'

'Did it work, Holas?'

'In my lifetime, Blue, yes, but perhaps I should ask you?'

'No more weirs of the kind you were talking about arrived in the Darling, Holas. But a huge weir went in near Menindee to hold back water in natural lakes for the whole Murray Darling system, to provide a more reliable supply for Broken Hill, and to save water for irrigation all the way down to the South Australian vineyards and orchards. Building started in 1949 and there have been upgrades ever since.'

'How did it influence navigation?'

'It stopped steamers going up river beyond Menindee, Holas.

But when the weir went in paddle steamers were no longer economical. Wool went out from the sheds on motor lorries.'

'There are no more paddle steamers?'

'Only for holiday joyriding on the Murray, Holas. Weirs and locks were built and assisted navigation on the Murray in the way you suggested, but there were none in the Darling.'

He paused and continued. 'Was my agitation wasted?'

'No Holas. In the time of paddle steamers you did influence people not to build weirs in the Darling and those built across the Murray helped to keep water in the Darling in the way you suggested.'

'Well that's good, Blue.'

'You were something of a formal letter writer, Holas, you gave lots of evidence at public enquiries and you presided at local courts too. Who did the work on the station?'

'Enough of that, Blue. You're starting to talk like Henry Lawson again. I had to cope with lots of changes at Albemarle as well as trying to keep governments on a straight path. Things were less profitable. Wool prices kept low following the depression and we had a problem getting rid of our surplus sheep. Cattle and horse markets dropped too.

The country was beginning to change. Rabbits were overrunning the place and we had a permanent system of poisoning them. It was costing us a big proportion of the wool clip each year to control them and the government of the colony of New South Wales helped with costs as well. We had two poison carts on the go all the time.'

'What was the poison, Holas?'

'Phosphorus, Blue. We mixed it with water, bisulphide of carbon and pollard and made it into tiny cakes. We used a two-horse buggy with two men, one driving and the other dropping the baits in a furrow the sulky made with a scarifier blade attached to the axle – rabbits always come to newly dug ground. We could lay baits in a 10-mile trail in a day – that way and we killed thousands of rabbits, but we never seemed to get on top of them. We had to learn to manage differently. There were so many new things we had to do. Lots of people were experimenting.

One experiment was with ways of preserving meat so we could ship it to Britain. Our Australian population wasn't growing as quickly as it had in the years following the gold rushes so we couldn't sell all the meat we had in the colonies. Great Britain wanted fresh meat. It wasn't able to produce all it needed and the North American colonies were able to ship it – especially in winter – they didn't need any refrigeration – their carcasses arrived fresh and sweet – naturally chilled along the way. We were too far away for that, and we had to go through the tropics. We had a few experiments with refrigeration and things were improving all the time but it was a difficult commercial proposition for us at Albemarle, so far from a seaport.

Canning meat to preserve it became popular, but even that was marginal. Some of it tasted vile. In the 1890s the best way to get a return from old sheep and cattle was to boil them down for their tallow and sell their skins or hides. A few of the boiling down works canned some of the meat as well but usually only luxury items went into cans.'

'What sort of luxury items do you mean, Holas?'

'Well the boiling down works at Menindee used to cook and can sheep tongues as a delicacy, and that was pretty convenient because that is how the slaughtermen were paid. Each tongue went into a box assigned to individual slaughtermen. The supervisors counted the tongues in each box, recorded the tally for each slaughterman and carried the tongues off to the cannery several times a day.

They tried all sorts of other things as well. There was an experienced food canner who even experimented with canning our beautiful Murray Cod and wild duck but I don't think they had a consistent enough supply to make the affair worthwhile. The boiling down works kept thousands of pigs and poultry. They lived on the offal from the sheep and cattle and they were slaughtered, dressed, cooked and canned. Some cans went off for export and others to Adelaide, Sydney or Melbourne. Someone even suggested that they preserve our native blackberry jam, but I don't think anything ever came of it.'

'Did the works never can carcass mutton?'

'I think it was difficult to separate the meat from the tallow in the digesters. The digesters were huge boilers – the tallow floated off with steam to wooden kegs, the remaining meat and bone was simply called gravy. When the pigs and poultry finished with it they dried it on the ground, scraped it up, crushed the bones for fertiliser, mixed it, bagged it, loaded it on steamers and sold the powder as manure to the fruit growers at Mildura. I can remember the manager boasting once that he was developing a demand from the sugar cane growers of Madagascar for his superior fertiliser.'

'It sounds like a vile place, Holas.'

'It was, Blue. Ladies on paddle steamers going past it used to hold their noses. The whole place was pretty filthy.'

'Was the works good for Albemarle?'

'My word, Blue. We used it from the minute it started in 1893. It was on the river and we could walk sheep in from the homestead in a day or so.'

'And the prices?'

'They varied with the seasons. We could pay the works a fee for processing our sheep – usually a shilling a head – and take the tallow and skins ourselves for sale in Melbourne or Adelaide, or the manager of the works would offer us a price per head.

In the centenary droughts at the end of the 1890s, we had to pay the boiling down works to take our starving sheep, or if we were lucky, we may have got a few pence for each. We had to get rid of them to get them off the country.

Overall, I think we averaged about 10,000 head a year going to the boiling-down works. But not towards the end of the century. Many of our stock had died, and those that were strong enough to travel had gone the boiling-down works already. By 1901 we had less than 19,000 sheep, a couple of hundred cattle and about 70 horses.

'How bad was that?'

'Terrible. For the first five years of the decade, we averaged about 110,000 sheep, 500 cattle and about 400 horses. We thought that was a good stable number, we had given up some country – we were down to about 700,000 acres, but every year after 1895 we got less rain than the year before. We struggled with increasing rabbits, and in 1898 we got the first of a series of four droughts in a row. It didn't rain again properly until 1903. By then we were down to less than 5,000 sheep with a handful of horses and cattle. We lost most of our stud animals – our special ram breeding flocks and nearly all our stud mares. The Darling went dry with just a few waterholes along it, there was no river traffic for years and many of our wells and dams dried up. We just kept waiting for rain. We felt hopeless.

It ruined the country. There were huge dust storms and we ended up with thousands of acres of clay pans that grew nothing because the topsoil was gone. Trees died. Kangaroos stopped breeding. Rabbits starved. Wells filled with sand. I didn't think anything would ever grow again in the way I remember it when I first came here to work with John and Joseph Phelps and Nicholas Chadwick. They would have been horrified to see the country they came to looking like that.

We had more Parliamentary enquiries. Nobody could make a go of the affairs they had with the rents we had to pay, but it was probably all too late. I think we had overstocked the country and the rabbits had probably made it twice as bad. The bare country encouraged a weed we called the Darling Pea. It poisoned horses. At the end of the drought, we had only two paddocks that were safe to keep horses in, but by then there were few left to poison.

The river was too dry for steamers and the tracks were too sandy for horse or bullock teams so we had to use camels to cart wool to Broken Hill and we mislaid a camel driver in a terrible dust storm. His name was Furarge and he got parted from the camel string – it was impossible to see more than a foot in front of your face. The other drivers didn't miss him for hours and when they went back to look for him they found no trace. This was Saturday. By the time I got two blacktrackers out to look for him, it was Monday and they didn't find him until Wednesday. He had water but he was without food for five days. Fortunately, Furarge seemed all right.'

Madam and the children came up for Christmas in 1896. From time to time Madam had a dry sense of humour. She had this temperature record put in one of the Adelaide newspapers:

The South Australian Register, Friday, 1 January 1897

A Warm and Cordial Christmas.— Mrs. N Sadleir has handed us the following record of the readings of the thermometer in the shade for Christmas week, taken at

Albemarle Station, New South Wales :— December 21, 105 ; 22, 105 ; 23, 115 ; 24, 114; 25. 118; 26, 114; 27, 106. Just fancy with such conditions prevailing, receiving from the old country a Christmas card with a beautifully chaste snowstorm scene thereon, 'wishing you a merry Christmas.' How delightfully ironical. Our correspondent also says a blinding dust storm blew continuously on December 25 and 26.'

'And what about the people at Albemarle, Holas?'

'It was terrible for everyone, Blue. Jack Phelps and I had the unpleasantness of putting people off because, with few stock, we had no work for them. We were able to promise them that when it rained again we wanted them back, but I felt terrible saying it. Some of the people who left had been born here.'

'Where did they go?'

'A whole contingent went to war from Albemarle. I think more than 20 of our young fellows enlisted in mounted units raised in New South Wales to go to South Africa to fight the Boers. Some of them did come back to Albemarle and it was delightful for me to greet them alive and fit. As far as I can remember, none of the Albemarle people died. If you think of it like that, the Albemarle fellows had a wonderful adventure – a sea voyage and the delights of South Africa while their old home waited for a good season to welcome them again. In many ways they did what they did here – riding, looking after horses, camping and shooting. They were made for the army in South Africa.'

'What did they shoot on Albemarle?'

'Wild dogs, rabbits, foxes and kangaroos. It was part of station hands' normal duties to control vermin.

A few more found jobs at the boiling-down works, and that was good, and we had some who went to White Cliffs looking for opal, but we never got back to the numbers of people we had before the Federation droughts. Menindee became a larger civilised settlement than Albemarle. Some of the people who worked here as wheelwrights, saddlers and harness-makers moved to town to set up as independent tradesmen. Jack Phelps and I helped them find premises to continue their trades and we patronised them when we could, but it was often only a gesture of friendship and kindness.

Albemarle was far less prosperous. We paid no profits to the estate of J.J. Phelps for years.'

'Did you maintain the gardens, Holas?'

'Yes. Two of the Chinamen stayed but we reduced the size of the vegetable gardens. There were far fewer people to feed. We kept most of the framework going – the store, the stables, the dairy and fowls, the outstations, the weekly mail and stores delivery buggy going as far as Booligal, the bookkeepers, the stable hands, the cook and butcher – and towards the end we had people carting water every day. It cost us a fortune in horse feed. We had used all our haystacks at Victoria Lake.'

'What about the Barkindji people?'

'In many ways the Barkindji came off worst of all. They were welcome on Albemarle, and there were families who remained there for all of my time, but many of them moved away to camps on the edge of settlements, at Pooncarie, Menindee, and Wilcannia – even at Ivanhoe away from the river. These were government ration camps. I don't know why our people went there. Their circumstances were more agreeable on Albemarle but I think families wanted to stay together and so they went off to join relatives in these camps. Some of the stations were less generous than we were and they didn't provide rations for the blackfellows and encouraged them to move to the ration camps. I reflected on it, and in many ways the Barkindji people lived and died like the labouring Irish class did in the workhouses of Cork and Tipperary I saw during the potato famine.'

'As bad as that?'

'Well yes. The Barkindji, on Albemarle at least, relied on their native food to supplement station rations for most of the time. They preferred it – kangaroo, wild duck, fish, echidna, goanna, snake, nardoo and so on. The continuous droughts got rid of all that. It was just like the potato famine at my old home. There was private charity but most food came from the government.'

11

Myths and Remains

The great-grandparents discuss myths about their lives and deaths. Nicholas asserts he lost his stations in a financial depression and banks took his funds – not drought. They respond to their great-grandson's description of the children's lives after their deaths. They learn four of their children spent some of their lives in Argentina and that four of their children served in the Great War. They describe weddings and family names and discuss myths and stories about their children. Great-Grandmother talks of her trip to Ireland and England and of her son James' visit to Dublin Castle to record the family's heraldic presence in Australia and they learn of James being offered a baronetcy when he was a bank manager at Loxton. They delight in their great-grandson's description and assessment of their children's success, but they dismiss him in ridicule when he describes the accoutrements of modern life in the 21st century.

I reflected on Nicholas Sadleir's pastoral life and matched it to family myths with the question: 'Holas, people said (I think I first heard it from my father) "Nicholas Sadleir walked away from 100,000 dying sheep in the droughts of the 1890s. He went broke. He managed stations in the Broken Hill district after that." What do you think of that?'

'Well history and families and myths are wonderful things, Blue. I think you know the tale isn't true, but I can explain some of it – or at least the reasons people believed "the walking away from 100,000 dying sheep" story – and so on. I think the main cause was Madam, and I may have contributed to it when we settled in Adelaide. Because of me, and because of Madam's pride, we talked about the stations I influenced as if we had a financial interest in all of them, and that was true for most.

By the time I gave up my interests in Queensland and north of Booligal, I was spending little time at Glenelg. We had emergencies at Albemarle and I needed to be there. Wool prices were dropping, I had less money to spend to maintain profits, rabbits were taking over the country and I had to put in new routines. I think Madam was content to say that I "was away attending to our pastoral interests" and that it was nobody's business to know that we no longer had any "pastoral interests" and I was simply managing a station for a family I had served for 43 years.

You could say that I walked away from 100,000 dying sheep in 1902 – I suppose it was metaphorically true – we had reduced our flocks at Albemarle by about 100,000. Not all of them starved, we were able to salvage some – but it was a financial disaster – and I shared some of its loss. I reduced my salary to £600 by agreement with Jack Phelps and his father Robert, who represented the estate, things got harder for Madam and the family, and Anna used the drought as a reason for our reduced circumstances. She loved poetry – "fortune being swept away by drought" had a lyrical ring to it and it seemed more tragically respectable than losing the trust of bankers in an economic slump. It also suited the story to support our selling the house at Glenelg and moving to a smaller place at the Grange.

Madam was a good trader. She sold the house at Glenelg at a profit, eliminated the mortgage, and bought cheaply into a block of newly built elegant terrace houses with fewer rooms, at the Grange, on the beach, facing the sea.

Madam was a good trader. She sold the house at Glenelg at a profit, eliminated the mortgage, and bought cheaply into a block of newly built elegant terrace houses with fewer rooms, at the Grange, on the beach, facing the sea. She saved face, we needed less space – some of the children were leaving home, and those remaining loved living on the beach. And she called it Ballysinode – after her ancestor John Hunt's estate in Tipperary. Gentility survived.

The children knew by then that the estate of J. J. Phelps owned Albemarle and some of them may have remembered Jack Phelps from Quamby days. Mary helped with the books and accounts when she was with me on holidays and so did de Vere, Jim, Toler and Robert from time to time, so they knew I didn't own Albemarle.

'Perhaps it was no mistake that Mary, James, de Vere, Toler, Jack and Kathleen chose careers in banking or with stock agents. You taught them accounting and the care of a large business.'

'Tell me some more about them, Blue.'

'James was a bank manager with the Bank of Adelaide.'

'Yes I remember him starting with the bank when he left school.'

'Well he opened branches at Tumby Bay, Streaky Bay and Loxton in South Australia and he managed the Yorketown branch. He married and had two children.'

'Good for Jim. Tell me about Mary. Surely she didn't become a bank manager. That would have been highly unusual.'

'When I met her as a little boy, Holas, she and Kathleen lived at the Grange. She was the most senior woman in the Adelaide office of Dalgety & Co. She never married. When Great-Grandmother died, she became the mother of the family.'

'And de Vere?'

'De Vere worked with the Bank of Australasia in Adelaide before he went to Argentina. He came back to South Australia from Argentina, married a girl from Yorke Peninsula, and took her back. I can remember Aunt Angela, who was with him in Argentina, telling me that he told her he was "a millionaire for a day". He was a skin dealer when that happened.

Robert went to Argentina as well, Holas.'

'Was he skin dealer too?'

'No. He went to Patagonia and started a sheep station. Argentinians call them estancias. You have many descendants there, and Sadleirs still breed Merinos in Patagonia.'

'Really? Something like Albemarle? But his mother wanted him to be a lawyer.'

'Why did he not?'

'Perhaps you should ask her.'

'Great-Grandmother, your son Robert went to Argentina, but his daughter, your granddaughter, believed that he graduated in law and was celebrating his graduation with an around-the-world tour before he settled there. Can you tell me anything about his legal studies or career?'

'A complex tale, Robert. Robert adored his father and as soon as he left Saints,[1] at the end of term just before Christmas, he went

---

1. St Peters School Collegiate - the first Anglican boys' school in Adelaide.

to work with him on Albemarle. He was an excellent scholar and I agreed to him going to Albemarle if he promised to return in March to apprentice himself as an attorney to his uncle Marshal at Mansfield in Victoria. I can remember the conversation at the railway station just before he left for Broken Hill.'

"Robert, enjoy your Christmas holiday, please give my love to your father, and remember I expect you home at Ballysinode in March. Your uncle Marshal expects you in Mansfield at the beginning of April, and we need to do some shopping to make sure you are properly turned out for a start in a new life. You will need warm clothes. Mansfield has cold winters."

"Yes, Mater," Robert said. "I promise. But please remember that Uncle Marshal has agreed to a trial only. I hardly know him. I respect your good opinion of me in believing that I can make an attorney, but I think you know that I would prefer to follow in my father's footsteps.'

I replied. "Yes, Robert, I know. You are as romantic as your father, but hardly a man of affairs. Did he ever tell you he could have been apprenticed as an attorney-at-law to associates of Sir Redmond Barry in Melbourne as soon as he came here from Ireland? Had he done so, I believe our fortunes would be better now."

Robert stared at the platform floor, reached up his face to kiss me, said "thank you, Mater," and boarded the train. He was a steadfast boy. I doubt I convinced him.'

'Did he go to Mansfield?'

Nicholas responded. 'He did Blue, but I had to work on him. It was in 1901. We were in terrible drought on Albemarle and I said, "Look around you, Robert. This is by no means beautiful. You have a wonderful opportunity to escape from it. I know you hark after this life and I, given my life to live again, would not change mine, but please take the opportunity your uncle offers. If, when you graduate, you still seek a life on the land and a good prospect offers, you may seize it. But wait. Finish your legal apprenticeship and see how things are."

'And what happened?'

Great Grandmother resumed. 'Robert seemed happy with his uncle and friends at Mansfield and then Marshal decided to close his practice because of ill-health – he died in 1903. Robert was somewhat stranded. He was able to resume his apprenticeship with other attorneys in Victoria – Marshal had arranged that – but we had no money to buy his indentures, so he came home to me at the Grange and went back to Albemarle with his father after staying a week or so. When Nicholas died at Albemarle, Robert was devastated. He and Elsie had nursed him and he had an expanded responsibility for the good conduct of the station, but he could see no part for himself in the new management and he came south to me after the funeral. Nicholas had left him a small amount of money and he spent it on a world sea voyage. Within a year he wrote to us from Argentina and soon afterwards de Vere joined him there.'

'So Robert followed in his father's footsteps after all.'

'Well we know he started to. Do you know how he fared, Robert?'

'Ronnie Land, his grandson, thinks he had a least 30,000 sheep on 300,000 acres in Patagonia. He married a South African and they had two sons and a daughter. The children went to school in Scotland and the daughter stayed there, married and had a daughter and two sons who remain in Great Britain. Robert's sons went back to Argentina after their schooling and carried on with sheep but I think they lost most of the business in an unsympathetic political regime.'

'I'm not sure what you mean, Robert.'

'I think they were taxed out of business and lost most of their lands.'

'Oh.' She was mute for a minute or two. She resumed with questions about de Vere.

'I remember de Vere coming back from the Argentine to marry Alice Waddell, a girl from Yorke Peninsula, a lovely girl, and we had a fine wedding at the Grange before they went to Argentina. We got one or two letters from him and Alice, but I died soon after they arrived in Argentina.'

'Well they had a son, and the son's widow lives in London, and de Vere lived and died in the province of Cordoba. Ronnie Land found shipping records of his imported and exported goods. When I was a little boy, Angela told me he had been a skin dealer. So I think he was an investor and a merchant.'

'We have not solved the puzzle of my children in Argentina, Robert. He said you thought that Charles went to Argentina. Are you sure he did?'

'Yes, Great-Grandmother. He enlisted in the British Army from there to serve in the Great War against Germany. De Vere enlisted too. They may have arrived in England to join Angela. She was nursing in London '

'Yes. We planned that. One of the reasons Angela and I went to England together was to install her in Guy's Hospital, London as a nurse in training. Angela was a sensitive girl. I encouraged her to study singing at the University of Adelaide to encourage a musical career, but she didn't really take to it. She seemed slightly aimless. I knew she was a caring girl and when I suggested training for nursing at Guy's hospital in London at the end of a tour together, she became enthusiastic.'

'There is another myth, Great-Grandmother. Barbara, your son Jim's daughter, believes that Angela saw a fortune teller who told her she could see her in institution dressed in white and Angela decided to go nursing because of it.'

'Well that may have helped. She never told me about it – which is a good thing because I may have disapproved of such pagan practices and discouraged her. I'm pleased things worked out for her. It wasn't planned that I die in Liverpool on the way back from Ireland, but at least I lived long enough to see some of the homes of the Hunts and the de Veres when I was there. I wasn't able to meet any of them, most of them had died, but Angela and I met many of Nicholas' relatives in Tipperary and some of my Sturgess relatives in England.'

I was lost for words. It seemed impudent to tell a dead person you were sorry she'd died. We communed in silence. Great-Grandmother resumed.

'And what was the Great War you spoke of Robert?'

'The Great War was terrible, Great-Grandmother. Millions of people died and it lasted five years. Great Britain and its Empire, France and Russia fought Germany, Austria, Hungary and The

Ottoman Empire. Federated Australia fought in it. Your four children served with British units.

'Goodness. Do you think de Vere left Alice, his wife, in Argentina to go off to fight on the other side of the world?'

'I suppose so, Great-Grandmother. He certainly returned to Argentina after the war, but Charles remained in Britain. Charles married Margaret Bobby and took up a farm in Scotland. As far as I know, he never returned to Argentina but he had a farm in Western Australia after Scotland. It was not a family success and his wife and sons returned to England. He served as a civilian in the Second World War in Australia and I think he ended his days as a mining prospector in the goldfields of Western Australia.'

Nicholas interrupted. 'That was Charlie. He was always looking for gold on Albemarle. Do you know if he found any, Blue?'

Great-Grandmother intervened. 'Gentlemen, we seem to have trouble keeping to the point. Goldfields are all very well, but how many more children of ours lived in Argentina, Robert?'

'My apologies, Great-Grandmother. It seems impossible to have sufficient knowledge about your issue's members. They formed an extremely interesting brood. To return to Argentina, Angela was the fourth. I think she went there for the first time after the Great War at her brothers' invitation. During the war she nursed in Great Britain, France and Italy. She was probably looking for a hospital with fewer dying young men. She worked as a nurse in Argentina for about 30 years. She returned to live at the Grange with Kathleen and Mary and ended her days in Adelaide.'

'She never married?'

'No. Georgina was the family's only bride and she died from maternal complications.[2] Perhaps she put the rest of the girls off.'

He resumed. 'Yes it was sad about Georgie. I think you've told me about her early death, Blue, but I'm interested in what you said about *four* of our children serving in the Great War. We talked about de Vere, Charlie and Angela. Who was the other?'

---

2. She died at 32. Her death certificate listed endometritis as the primary disease, and epileptic seizure and suffocation as the secondary disease. Probably she had been pregnant, had a miscarriage or a threatened miscarriage and died of 'complications' in hospital. There were no antibiotics then.

'It was John Raimond, your youngest son, Holas. He left the Bank of Adelaide at McLaren Vale in 1915 to travel to England to enlist in the King Edward Horse. He spent two and a half years in France and was invalided to Belfast to serve as an instructor. He never married. Sadly he died in a military hospital at Grantham in England of trench fever.'

'Trench fever, Blue?'

'Most of the Great War fighting happened in trenches dug in the ground. Soldiers lived in them in filthy conditions and many caught trench fever from the infective bites of body lice living in their clothing. Most people recovered, but it was a debilitating disease. Perhaps it wore Jack down and he died of something else. Did he have a bad heart?'

'No. He was a healthy, fit young chap as far as I can remember. Mind you, Blue, he was only 11 when I died. Madam, Jack had no ailments, did he?'

'No, Nicholas. He was a healthy boy and youth.'

'Where did the Raimond in his name come from Great-Grandmother? And why is it spelt like that? With an *i* instead of a *y*?'

'A Miss Raimond, a mature sensible and lovable woman served our households at Glenelg loyally from the time we arrived from Tasmania until after Kathleen was born. She helped at the birth of Toler, Tisha, Jack and Kath. She loved the children and they loved her and so we named Jack John Raimond. She was overjoyed. She burst into tears when we told her. Being a single woman she had no children of her own and Jack was her favourite forever. Sadly she died of a fever about a year after we moved into the new house at New Glenelg. We never had anyone like her again.

Isn't it strange? I never knew her Christian name. She was Miss Raimond to all of us, always.'

Great-Grandmother paused. 'I think I know why you're asking, Robert. You told me your father's name was Robert Raimond?'

'Yes.'

'And he was called Raimond?'

'Yes.'

'Georgie's first! Our first grandson! I held him as a baby.

Georgie was very fond of Miss Raimond too. We teased her about stealing Jack's name.'

'Miss Raimond must have been special, Great-Grandmother. What were her duties? Was she a nanny or a governess?'

'Neither, Robert. She was a housekeeper. She conducted the affairs of the house. If anyone was a nanny or governess it was me.'

'You ran a school at Quamby in Tasmania. Did you continue the tradition at Glenelg?'

'Well not in the way I did at Quamby. That was a proper licenced school. A superintendent of the Tasmanian Education Department inspected its conduct twice a year. I taught all my children at Glenelg but it supplemented their ordinary primary and secondary schooling. Sometimes I set them extra homework and I always helped them with their ordinary assignments from their schools.'

'Did any of the children go to university, Great-Grandmother?'

'Elsie studied mathematics and French. De Vere followed her as a non-graduating student and Angela studied music. I was hoping more of the children would attend and sit for bursary examinations (fees were out of the question) but they went to work. Elsie started a degree, but without a bursary she was unable to continue.'

'When did James go to England and Ireland, Great Grandmother?'

'He had been in the bank for several years and Nicholas helped him with his fare. Nicholas had always wanted to return to his home in Ireland but he was in the midst of a terrible drought at Albemarle and he thought that Jim, as the eldest son, should go to meet his cousins at Brookville House and register the presence of this Australian branch of the family at Dublin Castle. The bank gave Jim leave to go, but he delayed his departure after Nicholas's death. He went in 1907. I think he was about 31 when he got back and he opened a bank at Loxton. By then Tisha had left school and she went up to Loxton to work as his housekeeper.

'What did Dublin Castle have to do with the registration of families in Australia, Great-Grandmother?'

'It was where the Ulster King of Arms kept his records of noble families of Ireland. I encouraged James to have our lines of descent from the Sadleirs, and from my family, the DeVere Hunts, connected with the systems of heraldry and the records held at Dublin Castle.'

'Do you think James' registration at Dublin Castle connected with this event, Great-Grandmother?' And I read this newspaper article aloud.

The Register (Adelaide, SA) Wednesday 14 December 1904

Mrs. N. Sadleir, of Adelaide, received by the latest mail from England a copy of The Daily News containing notice of the death at Foynes Island, Co. Limerick, of her relative, Sir Stephen Edward De Vere, fourth baronet. Deceased, who was in his ninety-third year, was a brother of the Irish poet Aubrey De Vere, and was a prominent figure in Irish politics for upwards of half a century, representing Limerick County in Parliament for some years. He was a prolific writer, and his translations into English verse of the Odes of Horace are well known. The late Sir Edward De Vere was unmarried, and the baronetcy becomes extinct.

'Well yes, Robert. It was connected after a fashion. While Sir Stephen had no children there were other members of the family who did. It seemed to me that the Ulster King of Arms needed to be properly informed about possible lines of succession.'

'Yes, Great-Grandmother. Did you visit the office of the Ulster King of Arms when you visited Ireland with Angela?'

'No, Robert. There seemed little point. Jim told me what he had done and he reported on his conversations with some of the people in Dublin. He even had an official receipt for a payment he made for the registration of our family. By then Jim was the male heir of the family. It would not have done for me to meddle.'

'How did you spend your time in Ireland, Great-Grandmother? Who looked after the children when you left?

Anna Sadleir and daughter Angela at Dunloe, Killarney, Ireland.

'Well that was the wonder of it all, Robert. Charlie was in Western Australia; Tisha was with her brother Jim in Loxton, Kath and Toler were both working with Elder Smith & Co and Jack had just started as a junior with the Bank of Adelaide. The last of my little children were working and independent. They no longer needed me. I closed my Cellular Clothing Depot. I was free to have a holiday, and Angela came with me.'

'I think something came of James' dealing with Dublin Castle, Great-Grandmother.'

'Oh?'

'Here are two documents: A statement from Mary and Toler, and a newspaper article reporting your son James' death. I read them aloud.

COPY

About September, 1918, whilst at Loxton, S.A., staying with Jim I noticed a huge mail arrive for him, and observed that the large envelopes were addressed :-

Sir James Phelps Sadleir, Bart, J.P

Jim said that it was all a lot of rot, as far as he knew a Baronetcy could not pass through Mother to him, and in any case no money passed on with it, and he could not afford to keep it and, therefore he was returning all the <u>LETTERS PATENT & C.</u> by return mail, and on the envelopes had crossed out Sir and Bart stating that having been appointed a J.P. for the district he could not eliminate that.

He did remark to me that he had suggested that if they could send him along say, 8000 pounds p.a. he may be able to attempt to carry on the Baronetcy.

(Sgd.) A.T. Sadleir,

21/3/43

6,

Reginald Street,

Cott

W.A.

<u>Witness</u>
(Sgd.) Mary Sadleir
Grange,
S. Australia
21/3/43

NEWS. SATURDAY, JUNE 13, 1931
MR J.P. SADLEIR DEAD

One of the most remarkable figures in the South Australian banking, Mr James Phelps Sadleir, died suddenly at Yorketown yesterday afternoon. He was well-known to farmers.

<u>With his death was released a secret which he hid from the world. He was a baron by his own right, being the eldest son (in a family of 15) of the only daughter of a British Earl</u>. Mr Sadleir had been manager of the Yorketown branch of the bank of Adelaide since 1925.

For many years he suffered illness, but refused to leave his duties. He was at work yesterday morning.

Mr Sadleir was as much a bushman as a banker. He had a rough hearty personality which appealed to men out back. He was responsible for much bank pioneering work.

OPENED COUNTRY BANKS

Yorketown was the first country branch under his control which he himself had not founded. He opened branches at Tumby Bay in 1905 and Streaky Bay in 1906. Soon after his return from abroad in 1908 he began the Loxton office of the bank.

The banker would often set out into the bush and be away for days moving among the settlers, who cherished his acquaintance. He was born on May 21, 1876 and entered the service of the bank five days before his 16th birthday.

Mr Sadleir was the son of the late Mr Nicholas Sadleir of Albemarle station. Of his 15 brothers and sisters, Misses M, A G and K R Sadleir live in South Australia, Messers C H and A T in Western Australia, a brother and sister in South America, and a brother in Scotland.

He has left a widow and son and daughter.

Sgt R. A. Lenthall (Police Prosecutor) was stationed at Loxton and during that period Mr Sadleir managed the local branch. He made it his business to meet and help every settler in the district, said Mr Lenthall. He opened the bank in a tin shed and left it with a fine stone building and staff and offices. "He was a fine fellow, is respected and popular," concluded Mr Lenthall.'

Holas chuckled. 'Madam, I always imagined you beautiful, well-mannered, and finely bred, but never quite to the standard of *the only daughter of a British earl.*

There was a long pause. Great-Grandmother finally said, 'It is sad, Nicholas, that you greet news of your son and heirs' early demise with a form of mockery.'

'I am sorry, my dear. I fear I may have always been a practical man. I confess no surprise at Jim's early death. In fact, I am pleased he reached the age of 55. Do you remember his bout with Scarlatina at Quamby? Brother Richard told me he feared that Jim's heart was weakened when I took Jim to see him in Melbourne.'

I said, 'Perhaps I can help to explain things. I think for years people in South Australia believed the newspaper story. My mother, when she was engaged to marry my father, was teased by a stock agent who worked with Toler and Kathleen about my father's *blue blood*. It seems to me, it is another myth that needs explaining. Nearly two years ago, with Ronnie Land's help (Ronnie is your son Robert's grandson) I established that James Phelps Sadleir had no entitlement to any hereditary peerage (although I must hasten to add that there is no doubt that you, Great-Grandmother, and therefore I, are descended from the Earls of Oxford, and we are distantly related to the baronets Sir Aubrey and Sir Stephen de Vere of Limerick).

'That is as may be, but perhaps you and Ronnie as amateur genealogists are mistaken in comparison with the author of the LETTERS PATENT my son James received at Loxton?'

'I had my doubts as well, Great-Grandmother, so I employed a genealogist in Ireland who visited the Office of the Chief Herald and confirmed Ronnie's opinion (and even that of your son James). So I wrote to see if I could trace a copy of the letter in the Irish archives. I got nowhere with that and nor did the Dublin agent I employed. The records from that period simply do not exist and on reflection, I doubt if they ever did. I don't think the letter came from Dublin Castle.'

'Why?'

'Because in 1918 every one there – the administrative centre of British rule in Ireland – was preparing to resist an Irish rebellion against British rule in Ireland. The War of Independence did not begin until the following year but there was violence everywhere. I think the last thing any official of Dublin Castle would be doing was to offer an invalid baronetcy to a bank manager in Australia.

'It seems the Fenians never stopped Blue. Do you have any theories about where the letter came from?'

'I have only one that makes sense, Holas. Between 1916 and 1922, the British Prime Minister, with several assistants, offered hereditary titles and knighthoods to all and sundry for money. The Prime Minister's faction of the Liberal Party raised the money for its electoral campaigns. I think James may have been approached because of the interest in hereditary nobility he registered at Dublin Castle when he was there in 1907. I think those offering titles for sale may have looked at the files in Dublin Castle to decide who to write to.'

'But James didn't mention fees he had to pay. In his reply he suggested that the donors of the baronetcy may expect a favourable reply from him if the title came with £8000 a year.'

'Yes. But it is my guess that had your son James shown an interest in his reply to the LETTERS PATENT, then a bill would have been forthcoming.'

'Did many people buy titles?'

'It seems so, Holas. There was a huge increase in the number of titles awarded during the period, but in the end, the practice became illegal, and the payments kept secret so nobody really knew who had paid. It was said knighthoods cost £10,000, a baronetcy £30,000 and a peerage more than £50,000.

'Nicholas, had you not lost all that money in Queensland, perhaps Jim could have afforded it.'

'Madam!'

'A jest my love.'

'Jim did end up with a couple of farms in the Murray Mallee. Perhaps owning them compensated him for his lack of nobility, Great-Grandmother.'

Silence.

'I have one more myth to explore.'

'And?'

'All the surviving children were good at whatever they did. James has been mentioned. Mary was the most senior woman in Dalgety & Co in Adelaide. Kathleen was the same, but for Elder Smith and Company, and Toler managed the Western Australian division of it. Tish was an accomplished painter. Angela was matron of hospitals in Buenos Aires. Robert established a large estancia in Patagonia and de Vere was a successful trader in Argentina. Charles farmed in two countries. Even Jack and Georgina, who had shortened lives, lived fruitfully.'

'We accept your evidence immediately, Robert.'

'I concur, Blue. Have you any other myths for exploration?'

'Have I any other great-uncles or aunts?'

Silence.

'Did you ever find Helena?'

'No. We have no idea where she went despite all the energy we spent looking. We should have sought her earlier.'

'Have you any questions of me?'

Nicholas responded. 'You spoke of a war of independence in Ireland, Blue. Was it successfully repelled?'

'No, Holas. All the counties of Ireland except for most in Ulster comprise the Republic of Ireland. Ulster, except for Cavan, Donegal and Monaghan, remains part of Great Britain.'

Great-Grandmother responded. 'Is Australia a republic too?'

'No, Great-Grandmother. We still have the system of rule you remember after the federation of the Australian colonies. We have an English Queen – Elizabeth II – directly descended from your Queen Victoria and she is represented by governors in the states of Australia and by a Governor General for the Commonwealth of Australia. The Queen has a female Governor General and a female Prime Minister.

'Blue, I remember you telling me about the helicopters. You were joking when you told me that the new owners use them for mustering on parts of the old Mingara?'

'No, I was serious, Holas, it's true but I can understand why you thought I was telling tall stories. I imagine your father James in Tipperary would not have believed you if you told him you managed a station that shore at least 100,000 sheep annually with 40 shearers using mechanical shearing machines driven by a steam engine.'

'Yes. Madam and I saw many changes in our lives. What about your life, Blue?'

'Probably changes of equal importance, Holas – inventions and occupations you never heard of. You lived when the Telegraph started, but now it transmits the moving images and speech of real people on machines from any part of the world to another. Even the poorest people can afford to watch it in their homes. It is more popular than live theatre and largely replaces newspapers. People can see horse races, football and cricket without going to sporting grounds. They do it on machines, they carry in their pockets or bags, smaller than a deck of cards.'

'Thank you Robert. That is enough for today.'

We never spoke of it again.

Bibliography

Australian Council of National Trusts – Heritage Reprints, *Historic Homesteads,* Canberra, A.C.T, Australian Council of National Trusts – Heritage Reprints, 1982

Hardy, Bobbie, *Lament for the Barkindji: the vanished tribes of the Darling River Region,* Sydney, N.S.W.: Alpha Books, 1981

Lawson, Henry, *While the Billy Boils,* Australia: Angus and Robertson Publishers, 1896

Manning Clark, *A Short History of Australia, Second Illustrated Edition,* South Melbourne, The Macmillan Company, 1981

Marnane, Denis G, *Land and Violence – a history of West Tipperary from 1660:* D.G. Marnane, 1985

Mudie, Ian, *Riverboats,* Melbourne, Victoria, Australia.: Sun Books Pty Ltd. 1965

Richards – Mousley, Claudia, *Big Men Long Shadows: a story of the history and happenings of a sheep station on the River Darling – Windalle,* Kensington N.S.W.: Nelen Yubu Publications, 2010

Sadleir, John, *Recollections of a Victorian Police Officer,* Melbourne, Sydney, Adelaide, Brisbane and London: George Robertson & Company Propy Ltd, 1913

Sadleir, Richard M.F.S, *Five Centuries of Sadleirs – being a selected family genealogy – with interconnections to the families of Boyd, Clarke, Crofton, Grant, Hastie, Impie, Jones, Matthews, Rose, Sanderson and Tweedle:* unpublished, 2009

Webb, A, *The Progress of the Colony of Victoria,* Journal of the Dublin Statistical Society, Trinity College, Dublin, 1856

About the Author

Robert Hodge has had a lifetime career as an agricultural and farming consultant, since graduating in Agriculture from the Roseworthy Campus of the University of Adelaide, in 1961. He has worked for and consulted to Industry and Governments. As a trainer, lecturer, and Head of School he has taught and managed at various Australian and Philippines Colleges, and has advised on and managed State and Federal Government and Industry funded national and international projects.

He has lived and worked in Australia, Europe, North Africa, Asia, China and Laos. He is fluent in English and French and is a keen student of history.

Since retiring, he has travelled widely to research and write the story of a set of Australian great-grandparents. His recreational pastimes include wine-making from his small vineyard on the Fleurieu Peninsula, and drawing and painting. He is writing a novel on 21st-century migration to Australia.

See www,redhodgestories.com.au

www.ingramcontent.com/pod-product-compliance
Lightning Source LLC
Chambersburg PA
CBHW070754230426
43665CB00017B/2353